Cave Life

Cave Life *Evolution and Ecology*

David C. Culver

Harvard University Press
Cambridge, Massachusetts
and London, England
1982

Printed in the United States of America

Library of Congress Cataloging in Publication Data

Culver, David C., 1944–
Cave life.

Bibliography: p.
Includes index.
1. Cave ecology. 2. Population biology. I. Title.
QH541.4.C3C84 574.5′264 82-950
ISBN 0-674-10435-8 AACR2

To the memory of
Constance Campbell,
friend and colleague

Preface

This book is aimed at two audiences. First, it is for those interested in the role that theory does play and should play in evolutionary ecology. My own bias is that the fields of theoretical, laboratory, and field ecology should not be separate, and this book reflects that bias. The second audience that I hope will find this book useful is those interested in the natural history of cave organisms. In keeping with the theme of integrating various approaches to evolutionary ecology, I trust that readers primarily interested in theory and its applications will at least tolerate the discussions of natural history, and that readers interested in natural history will tolerate the discussions of theory.

Alfred Bögli, in the introduction to his book *Karst Hydrology and Physical Speleology,* says his interest in the hydrology of cave waters was crystallized when he was trapped by rising water in a cave for several days. Not having been trapped in a cave, I can only blame Kenneth A. Christiansen and Thomas L. Poulson for trapping me in the allure of caves and cave organisms. These two have pioneered a modern approach to cave biology, and both have made important steps in integrating biospeleology with the rest of ecology and evolutionary biology. It was my good fortune to have been a student of both.

Several colleagues provided preprints and unpublished material, including Kenneth Christiansen, John Cooper, Richard Franz, Janine Gibert, James Gooch, John Holsinger, Thomas Kane, Jacques Mathieu, and Thomas Poulson. I am especially grateful to John Cooper, Robert Mitchell, Arthur Palmer, and Stewart Peck for allowing me to use their photographs and drawings. Daniel Fong, John Holsinger, Thomas Poulson, and Judy Smallwood read part or all of the manuscript and made many useful comments. John Froeb did most of the illustrations, and Natalie Baker typed several drafts of the manuscript. Rosemary Grady helped with the manuscript in ways too numerous to mention. William Patrick and Peg Anderson of Harvard University Press provided constant encouragement and editorial assistance.

A large number of people have helped me with field work in caves over the years. I especially appreciate the help of Roger Baroody, Timothy Ehlinger, John Holsinger, and William K. Jones, all of whom helped out on numerous caving trips. Most of the caves I have visited are on private land, and I am grateful to the many landowners throughout the southern Appalachians who allowed me free access to their caves. Caving can be a dangerous undertaking. The National Speleological Society, Cave Avenue, Huntsville, Alabama, is a good source of information on caving techniques.

January 1982
Evanston, Illinois

Contents

Introduction

Cave biology is generally considered to be the study of extremes, often in the realm of scientific curiosity. The morphology of cave animals, with their reduced or absent eyes and greatly elongated appendages, is thought to be an extreme case of adaptation and an evolutionary dead end (Wake 1966). Specialization can be extreme, as in the case of seven species of Temnocephala that parasitize different body parts of one species of European cave shrimp, *Troglocaris schmidti* (Vandel 1964). Some cave animals seem "out of place"—the only known freshwater serpulid worm is from caves in Yugoslavia (Absolon and Hrabe 1930). Speciation can be extreme; there are over 150 species of beetles in the genus *Pseudanophthalmus* in North American caves (Barr 1969). This book, however, is not about extremes and curiosities, but about a very different topic—how cave faunas can help elucidate the general concepts and models of population biology, not because they are extreme, but because they are simple.

Modern population biology, in both its ecological and genetic sides, has come to have a strong mathematical base. Differential equation models of species interactions, optimization models of foraging behavior, and diffusion equation models of the fate of neutral mutations are only a few of many examples. There are several reasons why caves are useful places to test some of these models and concepts. First, compared to most habitats, selective pressures are simpler and easier to identify, scarcity of food and absence of light being the most obvious. Second, the relative stability of the cave environment makes it more likely that populations will be near equilibrium or steady state, which is an assumption of many models. Third, the number of species in many cave communities is small, which allows pairwise species interactions to be described more fully than is usually possible. Fourth, because of the large number of caves, natural replications are plentiful. There are over 2,300 caves in the state of Virginia, for example (Holsinger 1975), and caves are found throughout the world in both tropical and temperate areas (Fig. 1–1). Since the physical environments of caves at similar latitudes are very similar (temperatures are roughly constant and approximate mean annual surface temperatures), different caves are probably more similar than replicates in other habitats. Fifth, because of the simplicity of cave environments and cave communities, assumptions of many population biology models are more nearly approximated than in most other natural systems.

The very simplicity of caves might make them uninteresting to the population biologist. The prevailing way to test models has been to look at complex rather than simple communities, vertebrates rather than invertebrates, eucaryotes rather than procaryotes. However, optimism about our level of understanding of ecological and evolutionary processes is wearing very thin, and many models and concepts in ecology and population genetics are in doubt. One obvious example is the continuing controversy about the nature and even the amount of genetic variation in natural populations (Lewontin 1974). The inadequacy of the simplest single-locus selection models in accounting for observed variation has led to increased attention to neutral mutation models and to more complex selection models. But there are mathematical problems with the various statistical tests designed to determine if data are in agreement with prediction from neutral mutation theory (Ewens and Feldman 1976). Finally, most estimates of genetic variation may be too low (Coyne, Eanes, and Lewontin 1979) or too high (Finnerty and Johnson 1979), but in any case, they are incorrect.

One of the cornerstones of evolutionary biology is in doubt. A decade ago there was general consensus that speciation usually requires geographic isolation, but this consensus has broken down. Bush (1975) and Endler (1977) give convincing reasons why sympatric and parapatric speciation may actually be more common than allopatric speciation. The amount of genetic reorganization involved in speciation is not as great as previously thought (Ayala 1975).

Figure 1–1 Map of the karst areas of the United States. Caves also occur in some non-karst areas, especially in lava. (Map courtesy of Dr. Arthur N. Palmer, Department of Earth Science, State University College, Oneonta, New York.)

There are a multitude of other examples in ecology of once-accepted concepts that are now in doubt. Competition and predation may not be the only widespread interspecific interactions. The generally neglected phenomenon of mutualism is critically important, at least in some communities (Risch and Boucher 1976). Age-structure models are inadequate for many plant populations. Morphological stage may be more important than chronological age (Caswell and Werner 1978). Optimization models, which have particular importance in foraging theory and sex ratio theory, have been criticized on general evolutionary (Gould and Lewontin 1979) and experimental grounds (Herbers 1980).

There have been several responses to these problems. In some cases, it is reasonable to conclude that the phenomenon being studied is unimportant. For example, Strong, Szyska, and Simberloff (1979) have claimed that character displacement is not important in the classic case of the Galápagos finches. More usually, some combination of the model and the phenomenon being modeled is tested, with emphasis on the model. The island area effect is well documented, but it is unclear which mathematical model is best (Connor and McCoy 1979). One-locus models without selection cannot account for regressive evolution, but it is not clear whether multilocus models without selection are more appropriate or whether selection is involved.

Perhaps the most frequent result is that the model works in some cases but not in others. The most prominent example of this is in work on species interactions. Several investigators have asked whether differential equations of the form:

$$\frac{1}{N_i}\frac{dN_i}{dt} = a_i + \Sigma\, a_{ij}N_j \tag{1-1}$$

where N_i is population size, a_i is growth rate, and a_{ij}'s are interaction coefficients, are sufficient descriptions of the dynamics of a system. In some cases, they are (Vandermeer 1969), and in other cases they are not (Neill 1974). There has been a bewildering variety of responses to this situation, stemming partly from confusion over what questions are being asked. Is it that species interactions are critical or that linear terms of interspecific competition are critical? One response has been the argument that no nontrivial models in ecology are universal, and that falsifiability in the Popperian sense must be distinguished from usefulness, the correct prediction in some systems (Mankin et al. 1977). The danger here is the idea that the search for universal laws should be abandoned, a claim that usually predates the discovery of universals by a relatively short time. The history of physics is replete with examples of this.

Some researchers, using the Popperian doctrine of falsifiability, have claimed that a general model has been disproved, while in fact other assumptions were involved, for example, that the parameters of interaction, a_{ij}, were correctly measured. Too often there has been recourse to ill-defined concepts, such as "higher-order interactions," which measure our ignorance rather than our knowledge. Those problems have led some to reject a mathematical approach to certain phenomena (see Van Valen 1975). For ecologists this has meant an explicit rejection of any general theory. This is most clearly seen in much of systems ecology, where any correspondence between models of different ecosystems seems to come as a surprise. In population genetics, investigators have moved on to different problems.

What place do cave communities have in all this? I suggest the following modest, but potentially important, role. Because of their simplicity, cave communities in many cases are close to the assumptions of various ecological and evolutionary models. The environments are relatively constant, selective pressures are more or less identifiable, the amount of isolation is often known, and the entire set of potentially interacting species is known. For these reasons, models can be tested more completely, and the reasons for success and failure should be more transparent.

On the other hand, cave communities have important differences from laboratory systems. Cave populations have a long evolutionary and coevolutionary history, and they are more complicated than laboratory systems. For instance, in cave streams three species commonly interact, rather than two, as in most laboratory studies. This "moderate" complexity may point the way to understanding more complex communities.

This book attempts to show that cave biology can play a role in at least six areas of ecology and population genetics. Chapter 2 examines morphological, physiological, and behavioral adaptations to cave life and also serves as an introduction to the cave environment and some of its principal inhabitants. Because selective pressures are relatively simple, the results of adaptation and optimization should be easier to predict. Particular attention will be paid to alternative adaptive syndromes and possible cases of maladaptation.

In chapter 3, I review the evidence on life history characteristics of cave organisms. Since caves are generally food poor and environmentally stable, cave populations should have delayed reproduction and smaller clutch sizes. A starving population is likely to delay reproduction and reduce reproductive effort, so life history alterations can be either a direct consequence of or an evolutionary response to a food-poor environment. The emphasis will be on separating the two. Chapter 4 discusses regressive evolution in cave animals, focusing on whether mutation pressure can account for the rates of regressive evolution or whether selection must be involved.

Levels of allozyme variation within populations and between populations are summarized in chapter 5. With reduced environmental heterogeneity and generally small population sizes, genetic variation should be low. That this isn't the case for many cave populations sheds some light on the neutrality–selection controversy. Possible causes for the widely different levels of genetic differentiation among cave populations are examined. In chapter 6 I assess the validity of differential equation models of species interactions. Here, more than in any other area, there is positive feedback between models and their application to cave communities.

Chapter 7 clarifies the sometimes complementary, sometimes conflicting explanations of biogeographic patterns derived from historical and island biogeography. The question of the usefulness of the island analogy will be stressed.

This is not meant to exhaust the list of interesting topics in cave biology. The evolutionary history of cave organisms is a fascinating topic, but I have little to add to what has been said before. Likewise, I will only touch on our exploding knowledge of tropical cave faunas. Nor is this a catalog of cave fauna (see Vandel 1964), which badly needs updating, both for the United States and for the world. Similarly, this book is not a complete survey of models in evolutionary ecology. For some areas, such as life history models, even a review of the review articles would be quite lengthy. The models used in this book are, for the most part, simple and lie at the core of the questions being discussed.

Until very recently cave biology has been primarily descriptive, but there was an earlier attempt to give it a more central place in biology. In the latter part of the nineteenth and the early part of the twentieth century, cave biology played an important role in the rise of neo-Lamarckian ideas in North America (see Packard 1901). In the 1870s, fifteen papers on cave fauna were published in *The American Naturalist,* compared with the five papers on this subject in *The American Naturalist* in the 1970s. Many distinguished naturalists, including Eigenmann, Cope, Forbes, and Packard wrote about caves. For them, of course, the importance of cave organisms was that the loss of eyes seemed to support the theory of disuse. A. S. Packard summarized the arguments in *The Cave Fauna of North America with Remarks on the Brain and Origin of the Blind Species,* published in 1888. The book still makes interesting reading, and much of the best comparative anatomical work on cave organisms was done by neo-Lamarckians (see Eigenmann 1909). Cave biology did not play a direct role in the discrediting of neo-Lamarckian theories; this was a result of the rediscovery of Mendelian genetics and the demonstration by Fisher, Haldane, and Wright that Mendelian genetics could account for evolutionary change.

I do not expect that ecological and evolutionary theories will rise or fall because of data from cave biology, but the data can indicate serious problems for a theory, especially if special attention is paid to hidden assumption and conjectures. In the case of neo-Lamarckian theories, some assumptions were incorrect, and some data were, at the very least, difficult to explain. First, caves are much older than the several thousand years assumed by the neo-Lamarckians (Packard 1888). Second, loss of eyes takes much longer, at least in most cases, than the several generations they supposed. Third, not all cave-limited species are blind (see Besharse and Brandon 1973), as required by the theory of disuse. I will make every effort not to repeat that history and to consider data and theories as objectively and thoroughly as possible.

2 Adaptation

The study of adaptation is often a curious mixture of the obvious and the controversial. On the obvious side are observations that organisms living in a particular environment, say a desert, are better able to survive and reproduce in that environment than in environments where they do not naturally occur, say, a mesic forest. The controversies about adaptation are whether it is an optimization process (Lewontin 1979, Maynard Smith 1978), and whether it is universal. Adaptation and optimization require, first, the presence of genetic variation in fitness for the trait or traits being studied and, second, the absence of conflicting selective pressures or genetic constraints such as linkage that override or confuse the optimization process. For example, to state that the optimal solution to survival in the food-poor environment of caves is reduced metabolic rate (or increased metabolic efficiency) assumes that genetic variation in metabolic rate existed at some point and that there are no conflicting selective pressures, such as selection for rapid growth to escape predation.

Potential selective factors are often easier to identify in caves than in other environments. This chapter begins with a review of the abiotic and biotic conditions in caves that are likely to be important in selection. Then I will examine four predicted adaptations: the evolution of increased tactile and chemical senses to compensate for lack of light in caves, the evolution of metabolic economy, the evolution of adaptations to high moisture and humidity levels, and the evolution of neoteny. The focus throughout is on individual adaptation. The related question of the effect of individual fitness on population growth is discussed in chapter 3.

The Physical Environment

Since most biological work has been done in limestone caves in temperate climates, these are what I emphasize. It is worth noting that caves are not limited to limestone and are not limited to temperate regions. They occur in ice, such as Paradise Ice Cave on Mount Rainier, in tropical limestones, and in lava tubes throughout the world (Halliday 1974).

As a general rule, the physical environment of caves varies less than the surrounding surface environment. Physical conditions do vary, not only over time but also between caves and between areas within a cave. Relative to above-ground conditions, however, there is less variation below ground.

The temperature in caves approximates the mean annual temperature of the region. In parts of large caves remote from any entrance, the temperature scarcely varies at all. In remote parts of the Flint Ridge–Mammoth Cave system, air temperature varies between 13.6°C and 13.9°C (Barr and Kuehne 1971). In most caves, the temperature variation is greater. Figure 2–1 shows the annual temperature limits for Sainte-Catherine Cave in France. Such a temperature profile is typical of many small caves, with fluctuations of several degrees for most of the length and even greater variation near the entrance. Available evidence indicates that terrestrial cave organisms are very sensitive to variation in temperature. Juberthie (1969) found that the cave-limited beetle *Aphaenops cerberus bruneti* was present only at sites with low minute-to-minute fluctuations in temperature 2 mm above the soil (Fig. 2–2).

Relative humidity, even in dry passages, rarely falls below 80 percent. However, most terrestrial cave organisms are found in areas of near saturation. It is rare to find cave-limited species in areas that are not visibly damp or wet, and the terrestrial cave fauna is often restricted to the edges of streams and drip pools. Coping with high humidity and standing water may be a major feature of cave adaptation (see below and Howarth 1980).

Cave waters show a similar pattern of relative stability. Because of

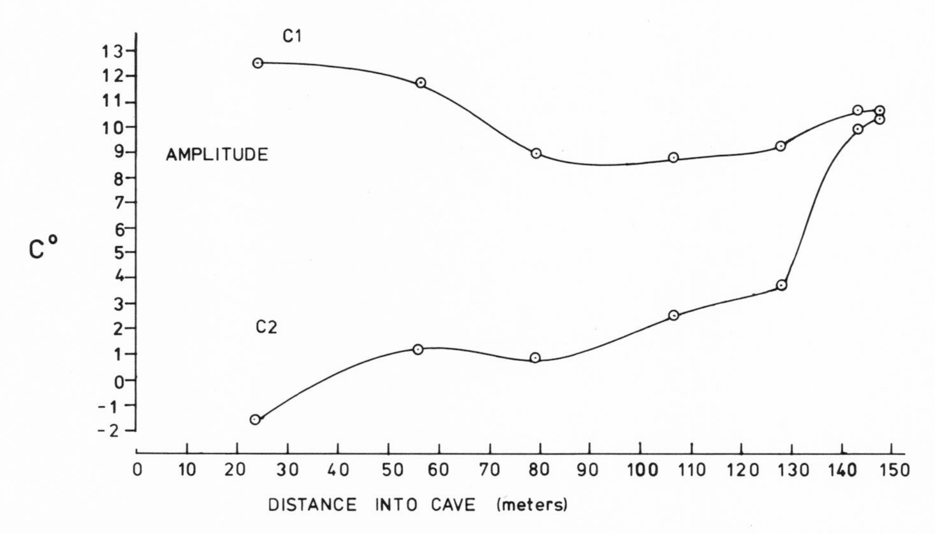

Figure 2–1 Annual temperature fluctuations throughout Sainte-Catherine Cave, Ariege, France. C1 is the maximum temperature profile, and C2 is the minimum temperature profile. (From Juberthie 1969.)

the porosity of karstified limestone, drainage basins in karst have moderate flood peaks compared to other drainage basins. Current velocity and discharge levels vary less in cave streams and in the rivers that are fed by cave waters as well. Pardé (cited in Jennings 1971) found that the mean monthly discharges of the River Nera, which drains a karst area, varied within only 17 percent of the mean annual discharge. By contrast, the Tiber varied up to 95 percent of the mean. On a smaller scale, White and Reich (1970) discovered that peak discharge rates were much lower for small karst streams than for nonkarst streams, but high discharge rates occurred over a longer time (Fig. 2–3). Many cave animals, however, are very sensitive to current flow. Spring floods are a major cause of mortality for cave amphipods (Culver 1971a), and reproduction is often cued to spring floods (Ginet 1960). Furthermore, caves with more variable hydrologic regimes have fewer species. In the Greenbrier Valley in West Virginia, caves with relatively mild hydrologic regimes had more cave-limited amphipod and isopod species ($\bar{x} = 2.3$, S.D. = 1.1, $n = 15$) than caves with severe floods that occasionally filled the cave passages ($\bar{x} = 0.8$, S.D. = 0.7, $n = 13$).

Chemically, cave waters are characterized by high alkaline hardness and relatively high pH (Table 2–1). Spring high water reduces alkaline hardness and pH, but both are still higher than in most surface streams.

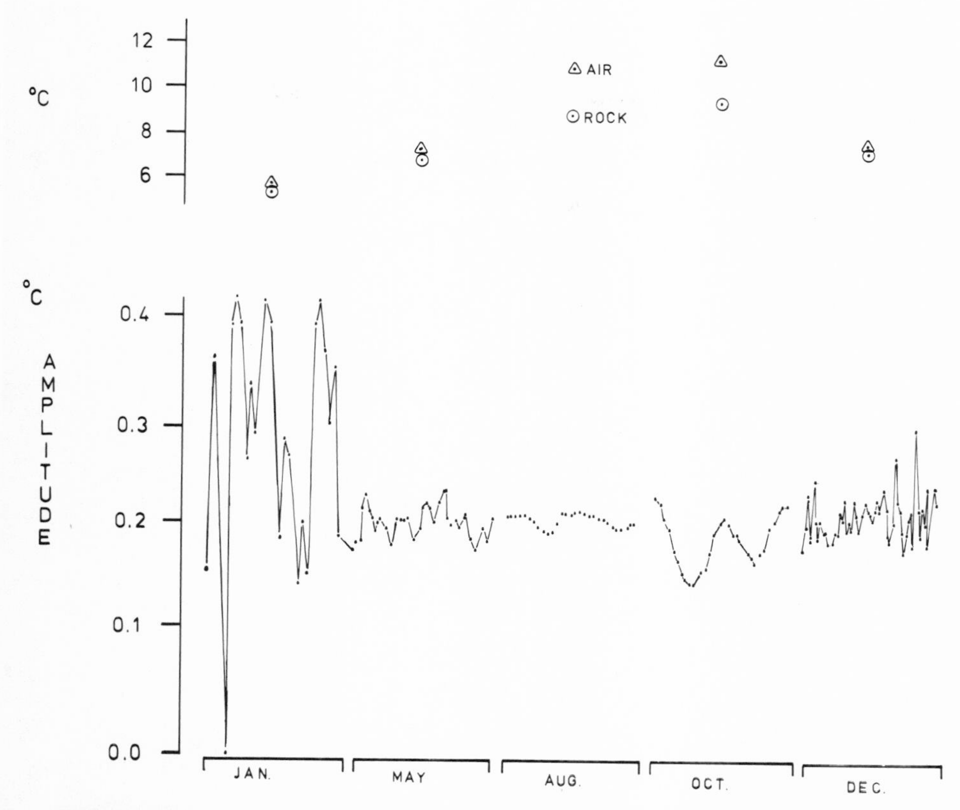

Figure 2–2 The effect of microclimate on the distribution of the beetle *Aphaenops cerberus bruneti* in Sainte-Catherine Cave. *Top panel,* mean temperature of air and rock. *Bottom panel,* deviation from the mean at 10-second intervals 2 mm above the soil. *Aphaenops* was absent in January and December when fluctuations were greatest. (Adapted from Juberthie 1969.)

In cave streams oxygen concentration is usually high, although in rimstone pools fed by slow drips and seeps it can be quite low. An extreme case is the water in Banner's Corner Cave in Virginia, which is polluted by septic tank leakage. Holsinger (1966) found pools there with less than 3 mg/l of oxygen. However, cave isopods (*Caecidotea recurvata*) and planarians (*Phagocata gracilis*) were very common.

Sources of Food

Except for a few chemosynthetic autotrophic bacteria that use iron and sulfur as an electron donor (Caumartin 1963), primary producers are absent. Thus, in a general sense, cave communities are decomposer

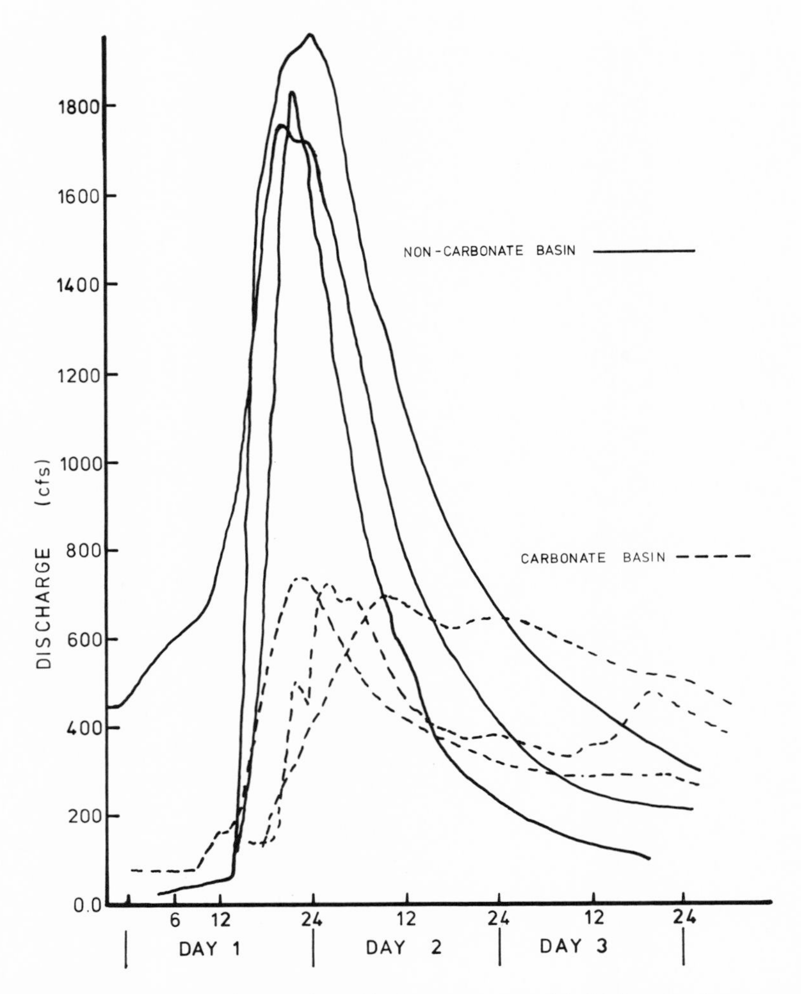

Figure 2–3 Peak discharges in cubic feet per second (cfs) for a carbonate and a noncarbonate basin in Pennsylvania for three days. The carbonate basin drains 145 km² and the noncarbonate basin drains 95 km². (From White and Reich 1970.)

communities. Food is brought in by both biological and physical agents continuously or in pulses and in different spatial configurations. Food enters a cave in three main ways. Organic matter is carried directly by streams and vertical shafts, and this source can be important for the terrestrial as well as the aquatic community because a layer of plant detritus is often left by receding flood waters. Another important source of food is the dissolved organic matter, bacteria, and protozoa in water

Table 2–1 Characteristics of limestone and sandstone waters. (From Sweeting 1973.)

Characteristic	Limestone, fast-flowing	Limestone, slow-flowing	Sandstone, fast- and slow-flowing
Total hardness	Low, less than 70, remains almost constant	High, increases with distance traveled, can be up to 250	Very low, 0–10
Alkaline hardness[a]	Low, less than 70, remains almost constant	High, increases with distance traveled, can be up to 250	Very low, often 0
Nonalkaline hardness	Low, remains almost constant	Low, remains almost constant	Often equal to total hardness
Calcium	Low, remains almost constant	High, increases with distance traveled, can be over 200	Very low
Magnesium	Low, remains almost constant	Low, remains almost constant	Very low
Free carbon dioxide	0–less than 10. More often 0	Varies considerably, 0–100	Varies, can be considerable
Acidity to methylorange	0	0	Can be considerable
pH	7.0–7.6, often constant	7.4–8.5	3.0–7.0
Color	Usually slightly colored	Colorless	Often highly colored

a. mg $CaCO_3$, per 1,000 ml.

percolating into the cave through the limestone rock. Finally, there are the feces of animals that regularly enter and leave the cave and eggs deposited by cave "crickets" (see below). In some caves, food may enter in more exotic ways. One most interesting special case is the Hawaiian lava tube caves, where exudates from tree roots are the major food source (Howarth 1972).

In the terrestrial cave environment there are at least five major sources of food: cave cricket eggs and guano, microorganisms, plant detritus left by flooding, bat guano, and feces of other mammals.

In many North American caves, cave crickets in the family Rhaphidophoridae, for example, *Euhadenoecus* and *Hadenoecus* in the east (Fig. 2–4) provide a major source of food. *Hadenoecus* leaves the cave

Figure 2–4 *Euhadenoecus fragilis* in Sweet Potato Cave, Lee County, Virginia. (Photo by author.)

at night and feeds "opportunistically and omnivorously as a scavenger" (Hubbell and Norton 1978), and cricket guano is an important food source for many species. Some of the most diverse terrestrial cave communities occur in areas where cricket guano is splattered on walls and floors. The female crickets oviposit inside the cave, usually in sandy substrates; in parts of the Edwards Plateau of Texas and the Interior Low Plateau in Kentucky, cave cricket eggs are the major dietary item for some beetle species (Mitchell 1968, Norton, Kane, and Poulson 1975). In some caves, this interaction comes close to being a naturally isolated predator–prey pair. In the absence of sandy substrates, the crickets oviposit in substrates that are difficult for beetles to excavate, and this interaction is absent.

Microorganisms occur on a variety of substrates, including wood, dung, and plant detritus, and are at least part of the diet of many terrestrial cave invertebrates. Microfungi appear to be more important than bacteria or actinomycetes. In a study of several Virginia caves, Dickson and Kirk (1976) found that the abundance of the cave-limited invertebrates was correlated with abundance of microfungi and with high fungal–bacterial ratios, but not with abundance of bacteria or ac-

Table 2–2 Average plate counts of microorganisms in and around Old Mill Cave, Virginia. (Modified from Dickson and Kirk 1976.)

Sample description	No. samples	Microfungi × 10^{-3} per gm of substrate	Ratio of bacteria: actinomycetes: fungi
A-horizon, forest soil	3	500	23:7:1
Entrance room	3	11	91:8:1
Dung, entrance room	1	7,160	6:0:1
Dark passages, floor	5	12	87:12:1
Floor with decayed chitin	1	490	25:0:1
Passages with stream	9	10	35:13:1
"Dry" passages	6	1	865:69:1

tinomycetes (Table 2–2). As expected, resource levels were lower in the cave than in forest soil, but there were exceptions. Dung in the entrance area and mud floors with chitin remains had high plate counts. The apparent correlation of high humidity and abundance of microfungi makes it difficult to know which factor is more important.

Plant detritus may also be an important food source. A layer of mud and finely divided leaves, often rich in oligochaetes, is deposited in many caves by slowly receding flood waters. Such areas often have a rich fauna. In caves subject to severe, rapid flooding, clumps of twigs and leaves are left behind, and the fauna on these resource patches is relatively distinct, usually of species not limited to caves, such as the isopod *Ligidium elrodii* (Schultz 1970, Holsinger, Baroody, and Culver 1975).

In stone-bottomed cave streams most of the food is plant detritus. Microorganisms are present, but in very low numbers. In slow-moving streams usually with mud bottoms, and in drip and seep pools isolated from the main stream, microfungi are more common and are correlated with the abundance of macroscopic invertebrates (Dickson and Kirk 1976). Plankton are scarce (Fig. 2–5), and are mostly washed into caves, but plankton populations apparently reproduce in the cave during the summer and autumn (Barr and Kuehne 1971).

A few caves harbor large bat colonies with large guano concentrations beneath, and in these caves food is comparatively abundant and continuously present. Thus selective pressures are different, and it is not surprising that the fauna feeding on guano is quite different from the rest of the cave fauna. Many guano-feeding species are not found in caves without guano, and these species display little of the eye and pig-

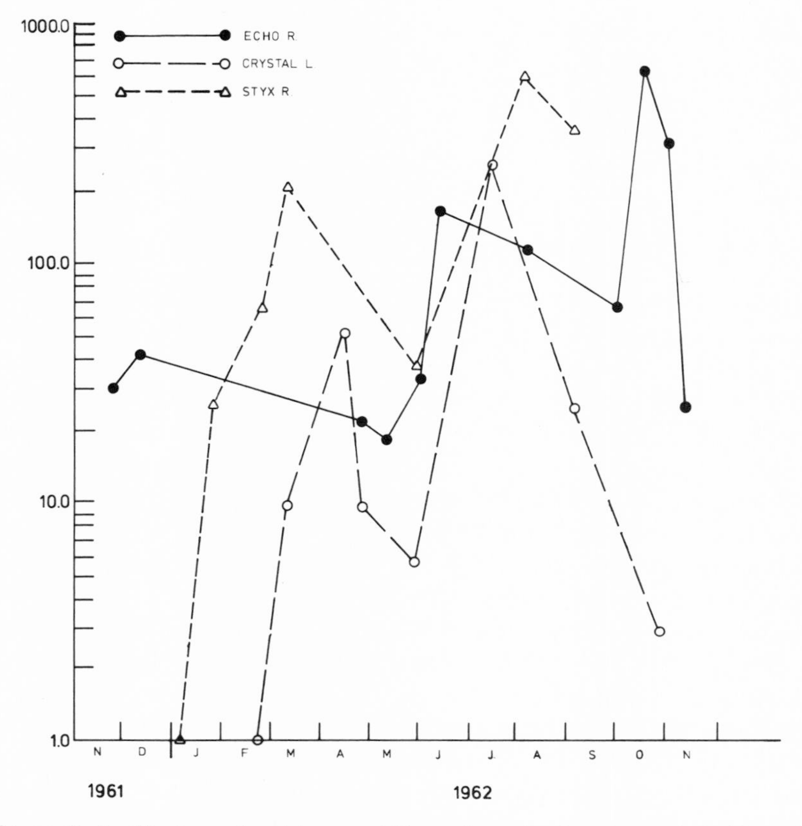

Figure 2–5 Plankton densities per 100 m of tow in three aquatic habitats in Mammoth Cave over a 12-month period. Peaks apparently correspond to heavy rainfall and spring snow melt. (From Barr and Kuehne 1971.)

ment reduction often found in cave organisms (Mitchell 1970, Peck 1971).

Besides serving as a substrate for microfungi, dung and dead animals are important food sources. Peck (1973a) has used dung as a very effective bait for catopid beetles. Poulson (1978) has found that there are differences in caloric content and in predictability of feces availability of mammals and invertebrates that enter caves, and that each fecal type has a more or less distinct community associated with it.

There remains the question of how much food is actually available to cave animals. The best comparative study is that of Peck and Richardson (1976), who compared the stomach contents of cave salamanders (*Eurycea lucifuga*) from the entrances and from the dark zones of caves in Tennessee and Alabama. Salamanders collected at the entrance, where one would expect food to be more abundant, had 18.6 prey items, with a volume of 0.14 ml, per stomach. By contrast, those from the dark zones had only 3.4 prey items, with a volume of 0.05 ml, per stomach.

Sensory Compensation

Perhaps the most cherished tenet of biospeleology, dating back at least to the neo-Lamarckians at the turn of the century, is that the fragile, delicate morphology of cave animals results from selection for increased sensory organs on appendages, which in turn results in lengthened appendages. This conjecture depends on two main points. The first is that the number (or size) of sense organs is increased by increasing the length of appendages bearing sense organs, or by increasing the density of sense organs per surface area, or by increasing surface area by appendage elongation. The second assumption is that increased number or size of sensory organs increases fitness. Since fitness is difficult to measure, it is not surprising that there is little evidence on this point. What is surprising is the paucity of evidence on the first point.

There are certainly cave organisms with very long appendages and a generally fragile appearance, of which *Euhadenoecus fragilis,* shown in Figure 2–4, is but one example. But how general is the phenomenon, and what is its adaptive significance? Morphometric studies are not that common in the biospeleological literature, and many suffer from inappropriate comparisons. For example, a comparison of European subterranean amphipods in the genus *Niphargus* with fresh- and brackish-water amphipods in the genus *Gammarus* (for example, Ginet 1960) is suspect, because *Niphargus* probably did not arise from *Gammarus,* and in fact the two genera are placed in separate families (Bousfield 1977). Inappropriate comparisons can work both ways. Albert Vandel, the noted French cave biologist, held to the outmoded orthogenetic theory that cave organisms are both phylogenetically and individually senescent (Vandel 1964) and therefore did not believe that adaptation to the cave environment occurs. Many of his often inappropriate comparisons of cave and surface populations show no lengthening of appendages of cave populations (p. 572).

To investigate whether appendages of cave organisms are indeed longer, two comparisons must be made. First, the appendage lengths of cave populations should be compared with those of their closest noncave relatives. But since the ancestors of cave organisms often occur in habitats sharing some characteristics of caves, especially reduced light levels, they are likely to be "preadapted" to cave life. The second comparison should be between the appendage lengths of the ancestors or relatives of cave species and those of other noncave populations. Such a morphometric study of preadaptation has not been attempted and indeed would be very difficult because of the very large number of measurements required. Morphometric studies of cave and related noncave populations have been made, but most of these studies have involved only adults, and little information on ontogeny is available, making it difficult to interpret the allometry equation $y = bx^a$ (Gould 1971).

The amphipod *Gammarus minus* is common in springs and caves in much of the Appalachians. Populations in the large caves of Greenbrier County, West Virginia, and Tazewell County, Virginia, are noticeably paler and longer, with reduced eyes and elongated appendages. A reanalysis of data for these populations given by Holsinger and Culver (1970) shows that the increase in length of the first antennae in cave populations is not due to positive allometric effects, where an increase in body size causes a proportionally greater increase in antennal length. Fits of both the large cave and the spring populations to the allometry equation $y = bx^a$, where x is length and y is antennal length, yielded values of a slightly less than, but not significantly different from, 1. The linear slopes are significantly different (0.52 for springs, 0.71 for large caves), resulting in quite different morphologies (Fig. 2–6). The large cave forms also show considerable eye degeneration, and in fact Bousfield (1958) suggested that the large cave form might be a separate species. These morphometric changes would seem to support the claim that increased appendage length is a result of adaptation to the cave environment. However, the genetic basis of these changes is unclear. In an analysis of electrophoretic variants of soluble enzymes, Gooch and Hetrick (1979) found considerable geographic variation, but almost no differences between spring and cave populations. This suggests that adaptation to cave life does not require a large reorganization of the genome. Finally, there is no direct evidence that the morphometric changes observed are the result of selection or that they even have a genetic basis.

Morphological studies of other cave organisms have yielded a variety of results. An analysis of Shear's (1972) data for type specimens of species in the milliped genus *Pseudotremia* indicates that the cave-limited species do not have proportionately longer third antennal segments than other species (Table 2–3). Using the allometry equation

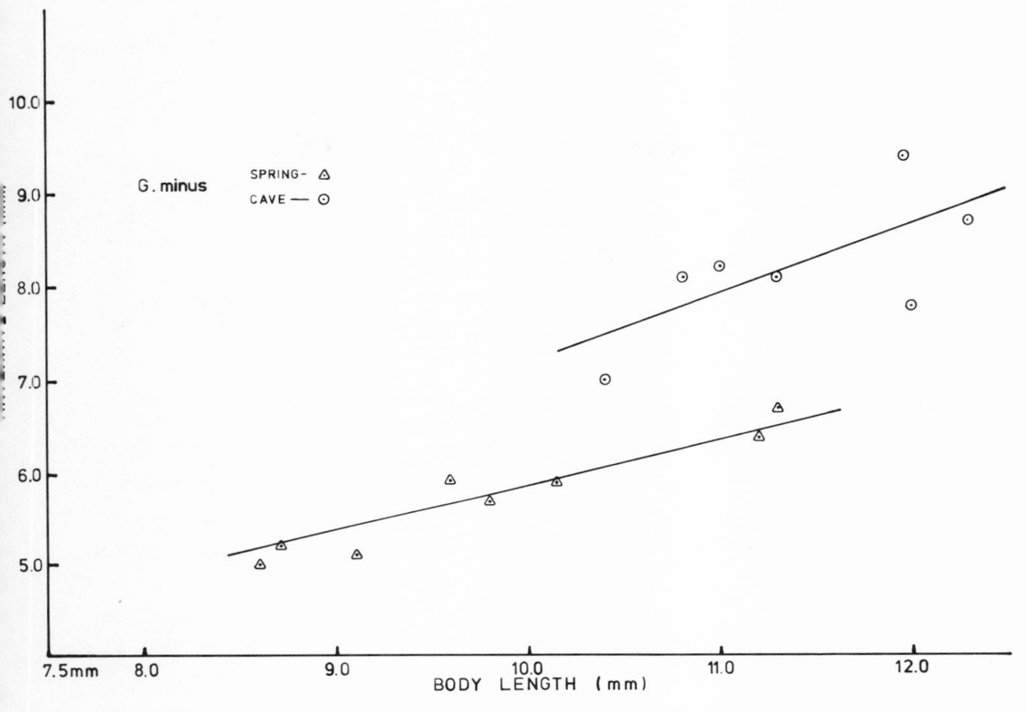

Figure 2–6 Average body length plotted against length of first antennae in mature male *Gammarus minus* from eight spring and seven cave populations in Virginia and West Virginia. (Modified from Holsinger and Culver 1970.)

$y = bx^a$, neither b nor a was consistently higher for cave-limited species. The only group that showed consistent differences were cave-limited species that retained pigment. They had a lower slope, but a higher intercept than either surface species or cave-limited species lacking pigment.

Peck's (1973a) study of *Ptomaphagus* beetles, on the other hand, indicates that cave species do have elongated antennae. Although he only gives ratios or drawings of antennae, making allometric analysis impossible, it is clear from his drawings that the cave species have longer antennae. Peck suggests that long antennae are advantageous to cave species because they allow greater searching ability and disadvantageous to surface species because space is a constraint in their soil-litter habitats. Antennae of Mexican *Ptomaphagus* are shown in Figure 2–7. Most of the cave-limited *Ptomaphagus* have antennae like *P. troglomexicanus,* but a small group of species that are regularly found

Table 2–3 Morphometric analysis of the relationship of length of the third antennal segment to body length of type specimens of the milliped genus *Pseudotremia*. Data were fitted to the allometric equation log y = log b + a log x (compare $y = bx^a$), where x is body length and y is third antennal segment length.

Category	a	s_a	log b	$s_{\log b}$	No. species
Without pigment, cave-limited	0.67	0.20	−0.81	0.24	15
With pigment, cave-limited	0.35	0.20	−0.38	0.26	8
With pigment, cave and surface habitats	0.80	0.14	−0.99	0.21	8
With pigment, surface-limited	0.61	0.30	−0.77	0.43	5

in caves have short antennae like *P. elabra*. Peck follows accepted procedure when he divides those species known only from caves into two groups. The first, larger group consists of those cave-limited species with significant eye and pigment reduction. The second, smaller group consists of species known only from caves but with little eye and pigment reduction. The first clearly are troglobites—obligate cave-dwelling species. The second are usually called troglophiles—facultative cave-dwelling species—because on the basis of their morphology, it is assumed that they will be found in noncave habitats.

It is also widely held that gene exchange with surface populations prevents troglophiles from fully adapting to the cave environment. There is an obvious circularity in this argument, because species that do not show appendage elongation or regressive evolution are assumed to be swamped by gene flow or to have not been in caves long enough to adapt (see Hamilton-Smith 1971). However, many of these troglophiles are very successful in caves, at least numerically. The best examples are the little-studied cave Diptera, which are common in many caves. Gene flow and insufficient time may have constrained the amount of appendage elongation, but it is also possible that not all cave species are subject to selection pressures for sensory compensation and appendage elongation. Some credence is given to this view by Franz and Lee's (in press) suggestion that cave-limited crayfish are adapted to different resource levels, and by Poulson's (1978) data showing that some less modified cave species specialize on locally abundant but patchily distributed resources. The most parsimonious explanation of the morphometric data is that some cave populations undergo considerable appendage elongation, probably the result of selection for increased nonvisual sense organs in the food-poor environ-

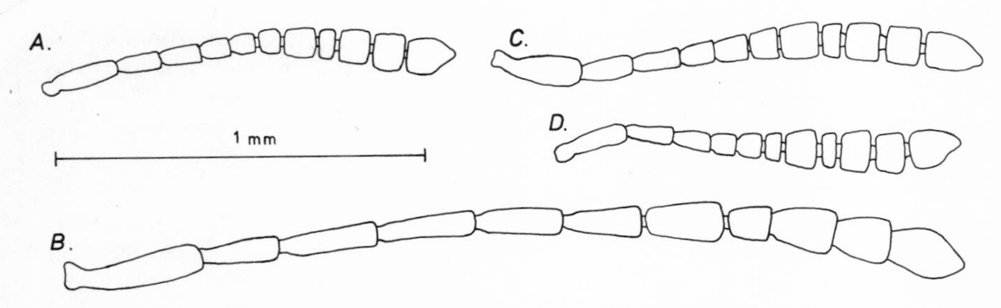

Figure 2–7 Antennae of Mexican *Ptomaphagus* beetles: (*A*) *P. elabra*, a cave-limited species with large eyes; (*B*) *P. troglomexicanus*, also cave-limited; (*C*) *P. altus*, found only in forests; and (*D*) *P. leo*, found in caves and in forests. (From Peck 1973a.)

ment. Other cave populations do not, and it is at least possible that appendage elongation is not being selected for.

One especially clear-cut example of the evolution of increased sense organs in cave species is given in Cooper's (1969) study of cave and surface *Orconectes* crayfish. She minimized allometric effects by comparing crayfish of the same size. As indicated in Table 2–4, the cave-limited *Orconectes australis* has longer antennae and antennular flagellae, more antennular asthetascs, and longer asthetascs than the surface-dwelling *O. limosus*, the closest surface relative to *O. australis*. Apparently both chemoreceptors (asthetascs on the external flagellum) and mechanoreceptors (various organs on the internal flagellum and second antenna) are used to locate food. In a series of laboratory experiments Cooper showed that *O. inermis*, a cave-limited species very similar to *O. australis*, located injured *Enchytraeus* worms twice as fast as *O. limosus* whose eyes had been painted with enamel.

Vandel (1960) has provided counterexamples of evolution of decreased sense organs in cave species. Eyeless, unpigmented cave-dwelling isopods in the genus *Androniscus* have fewer antennular asthetascs than the species Vandel believed closest in morphology to the ancestral stock, *A. dentiger*, even though the cave-limited species have the same body length. Vandel's interpretation is that these data support the orthogenetic hypothesis of phyletic senescence of cave species and their lack of adaptation. A more modern interpretation is that as with *Pseudotremia*, discussed above (see Table 2–3), there has not been selection for appendage elongation and increased sense organs in cave-dwelling species. That antennular asthetascs are actually reduced

Table 2–4 Comparison of antennular and second antennal morphology of male *Orconectes australis,* a cave-limited species, and male *O. limosus,* the closest surface relative of *O. australis.* (Data from Cooper 1969.)

Characteristic	*O. limosus*	*O. australis*
Carapace length	30.5 mm	30.5 mm
Second antenna length	42 mm	86 mm
Internal antennular flagella length	9.0 mm	12.2 mm
External antennular flagella		
Length	10.5 mm	11.4 mm
No. segments	31	36
No. bearing asthetascs	18	22
No. asthetascs/segment	7	7
Asthetasc length	0.12 mm	0.18 mm

in cave-limited *Androniscus* is open to question. All species in the subgenus *Dentigeroniscus* have three antennular asthetascs except for *A. dentiger.* All species, including *A. dentiger,* are found in caves, but some, for example, *A. brentanus* and *A. dentiger,* are found outside caves as well. Furthermore, other sense organs may display different patterns.

The classic case of sensory compensation and cave adaptation in general is Eigenmann's (1909) and Poulson's (1963) studies of amblyopsid fish. Since these studies have become a paradigm, they are worth examining in some detail. At the outset I should point out that cave amblyopsids are the top predators in the aquatic cave food chain, and to a degree that is extreme even for cave populations their food is both diffuse and scarce. It presents a most likely case to look for sensory compensation.

The Amblyopsidae comprise six species in four genera. *Chologaster cornuta* is nocturnal and lives in swamps in the Atlantic coastal plain; *C. agassizi* is found in springs and caves, more commonly in springs; and the remaining four species are limited to caves. Assuming no differences in the rates of regressive evolution among the species, it is possible to rank these four according to the length of time they have been in caves by comparing relative eye and pigment cell degeneration (Poulson and White 1969, Cooper and Kuehne 1974). *Typhlichthys subterraneus* shows the least regressive evolution, followed by *Amblyopsis spelaea, A. rosae,* and *Speoplatyrhinus poulsoni.* There appear to be two main phyletic lines (Woods and Inger 1957): the *Chologaster–Typhlichthys* line and the *Amblyopsis* line. The anatomi-

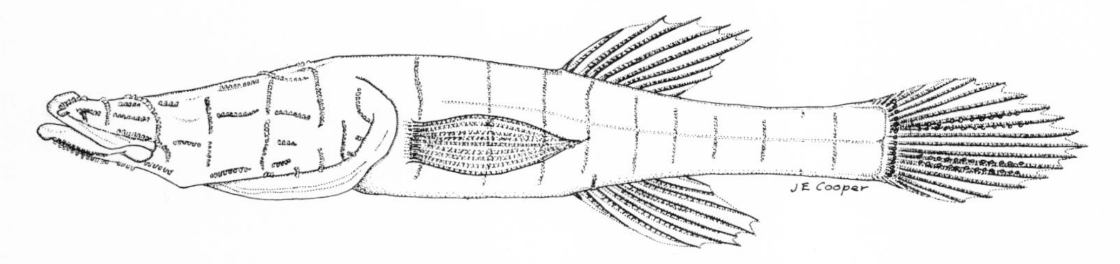

Figure 2–8 The cave fish *Speoplatyrhinus poulsoni*. (Drawing by Dr. John E. Cooper, courtesy North Carolina State Museum of Natural History.)

cal affinities of *Speoplatyrhinus* are obscure, suggesting a third line; this remarkable fish is shown in Figure 2–8.

The most striking sensory hypertrophy is in the lateral line system, shown in Figure 2–9. This hypertrophy appears to increase food-finding ability at low, but not at high, prey densities. When one *Daphnia* was introduced into a 100-liter aquarium, *A. spelaea* found it hours before *C. agassizi* did (Poulson and White 1969). In contrast *C. agassizi* ate all ten *Daphnia* introduced in a 5-liter aquarium before *A. spelaea* had eaten half. The results at high prey densities probably result from a low maximum food intake of *Amblyopsis*. Table 2–5 shows reaction distances of the amblyopsids to various prey. Because of differences in neuromast morphology, *Typhlichthys* is more sensitive to general water movement, and *Amblyopsis* is more sensitive to direction (Poulson 1963).

Brain anatomy also undergoes adaptive changes with increasing time of isolation in caves (Fig. 2–10). The optic lobe decreases in size, but more interesting are the increases in size of the telencephalon, cerebellum, semicircular canals (dynamic equilibrium receptors), otoliths (static equilibrium receptors), eminentia granularis (input path for lateral line and semicircular canal–otolith input), and cristae cerebellum lobe (input path for tactile receptors) (Poulson 1963). The only exception to this trend is the smaller cerebellum size in *Speoplatyrhinus* (Cooper and Kuehne 1974). Obstacle avoidance and spatial memory are increased in the cave-limited species, especially *Amblyopsis,* as a result of these changes.

Since there has been selection for food-finding ability because of a food-scarce environment, one would also expect selection for increased metabolic efficiency in these species, and in fact this is the case. All three cave-limited species have metabolic rates considerably

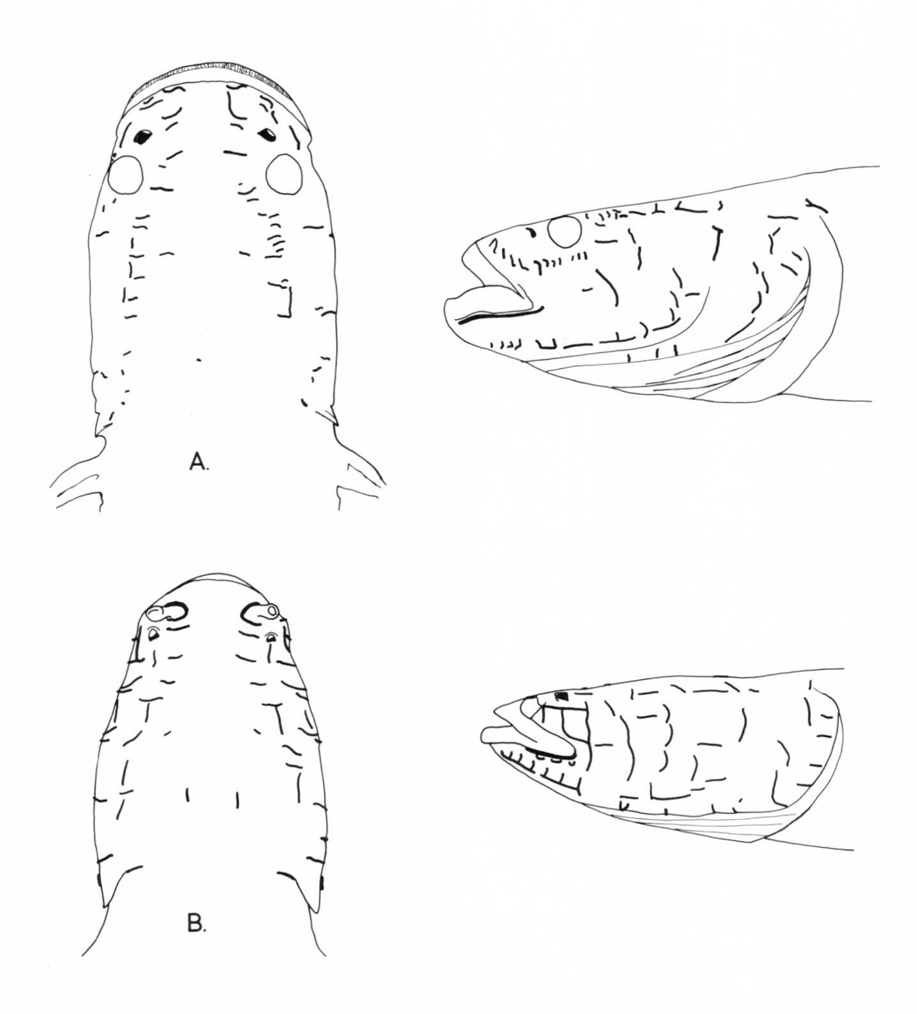

Figure 2–9 Dorsal and lateral views of the head lateral line system of (*A*) *Chologaster agassizi*, (*B*) *Typhlichthys subterraneus*, (*C*) *Amblyopsis spelaea*, (*D*) *A. rosae*, and (*E*) *Speoplatyrhinus poulsoni*. (Modified from Eigenmann 1909, Poulson 1963, and Cooper and Kuehne 1974.)

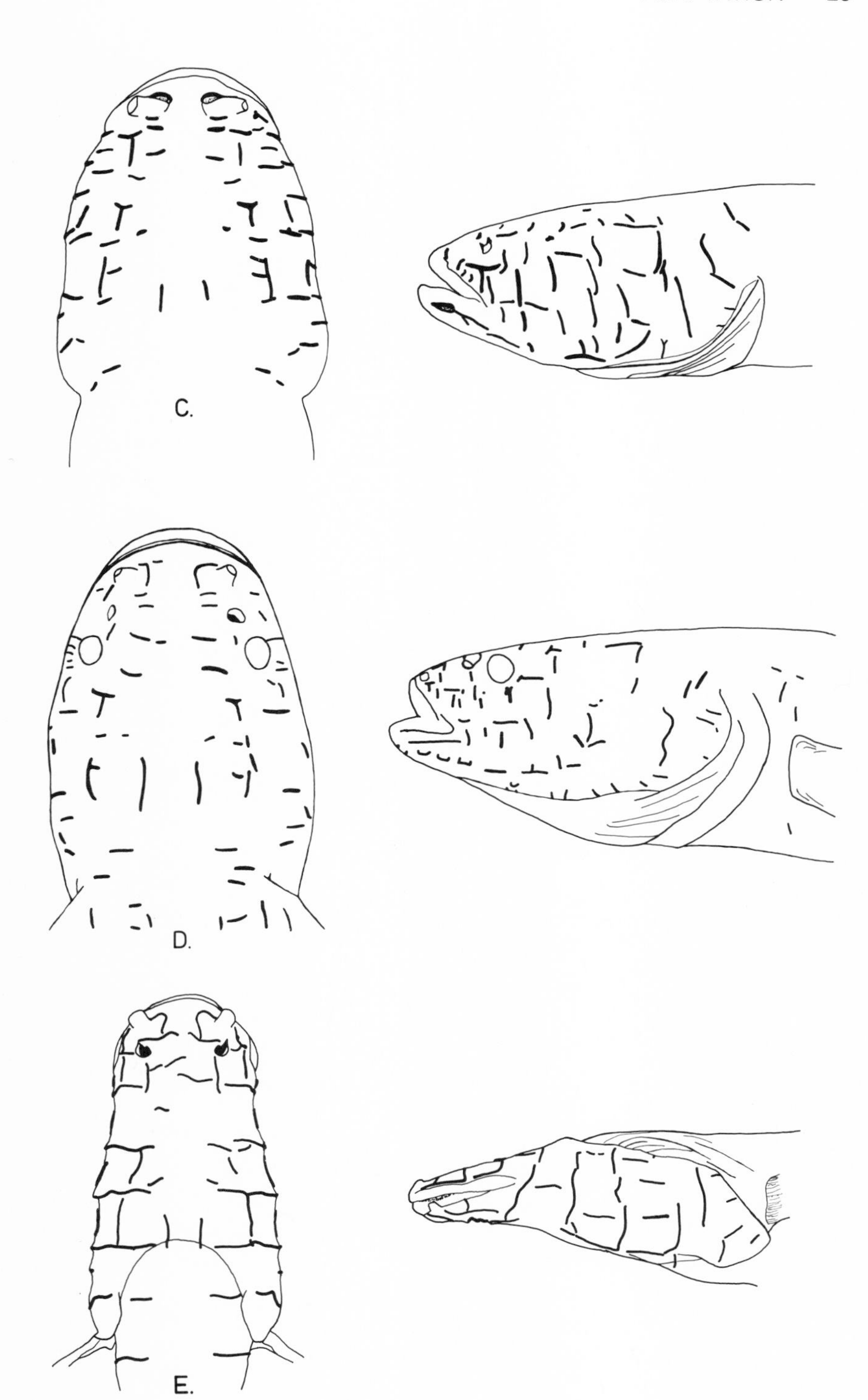
C.
D.
E.

Table 2–5 Maximum distance of orientation toward prey of various amblyopsid fish. (From Poulson 1963.)

	Prey		
Predator	*Eubranchippus*	*Daphnia*	*Hyalella*
Chologaster cornuta	20 mm	—	—
Chologaster agassizi	—	10 mm	20–30 mm
Typhlichthys subterraneus	—	30–40 mm	20–50 mm
Amblyopsis spelaea	—	20–30 mm	30–45 mm
Amblyopsis rosae	—	30–45 mm	—

lower than those of the spring-dwelling *Chologaster agassizi* (Poulson 1963); see Fig. 2–11.

Metabolic Economy

If food is generally scarce in caves, one might expect selection for reduced metabolic rate, because a lower standard metabolic rate (SMR) would allow greater resistance to starvation and because a lower routine metabolic rate (RMR) would make more energy available for reproduction. Most of the studies, aside from Poulson's, suffer from one or more defects (comparison with inappropriate surface species; failure to control weight, activity, or temperature), but a general pattern emerges: usually, *but not always,* cave species have low metabolic rates.

First let us consider those cases that agree with the amblyopsid paradigm. Cave crayfish show low metabolic rates; the cave-limited *Cambarus setosus* survives 3.5 times longer than the surface-dwelling *C. rusticus* on the same amount of oxygen (Burbanck, Edwards, and Burbanck 1948). Dickson and Franz (1980) showed that cave-limited species of *Procambarus* had lower gill respiration rates and longer adenosine triphosphate (ATP) turnover times than surface-dwelling species. They further showed that differences between the two cave-limited species studied were correlated with food availability. *Procambarus franzi,* taken from a cave with a large bat roost and hence high organic input, had higher gill respiration rates and shorter ATP turnover times than *P. pallidus* from a cave with low organic input. This contrasts with Poulson's amblyopsid study, where differences in metabolic rate appeared to be related to differences in the length of time isolated in caves rather than to differences in resource levels. One of the

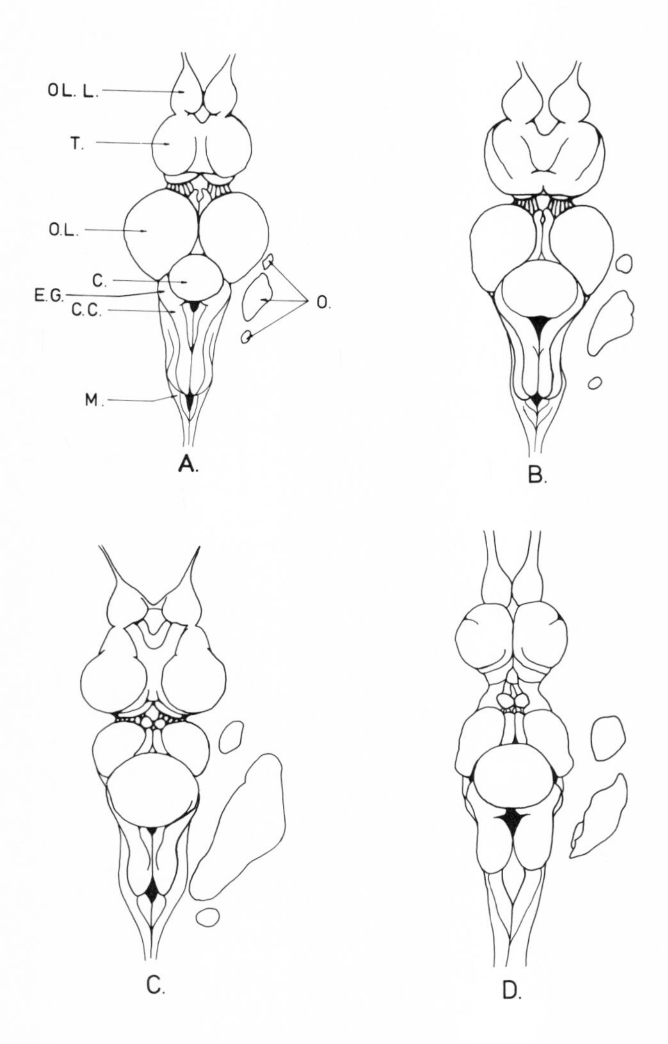

Figure 2–10 Brain morphologies of (*A*) *Chologaster cornuta,* a surface species; (*B*) *C. agassizi,* a troglophile; (*C*) *Amblyopsis rosae,* a troglobite; and (*D*) *Speoplatyrhinus poulsoni,* a troglobite. Parts labeled are: OL.L., olfactory lobe; T., telencephalon; O.L., optic lobe; C., cerebellum; E.G., eminentia granularis; C.C., cristae cerebelli; M., medulla oblongata; O., otoliths. (From Poulson 1963, and Cooper and Kuehne 1974.)

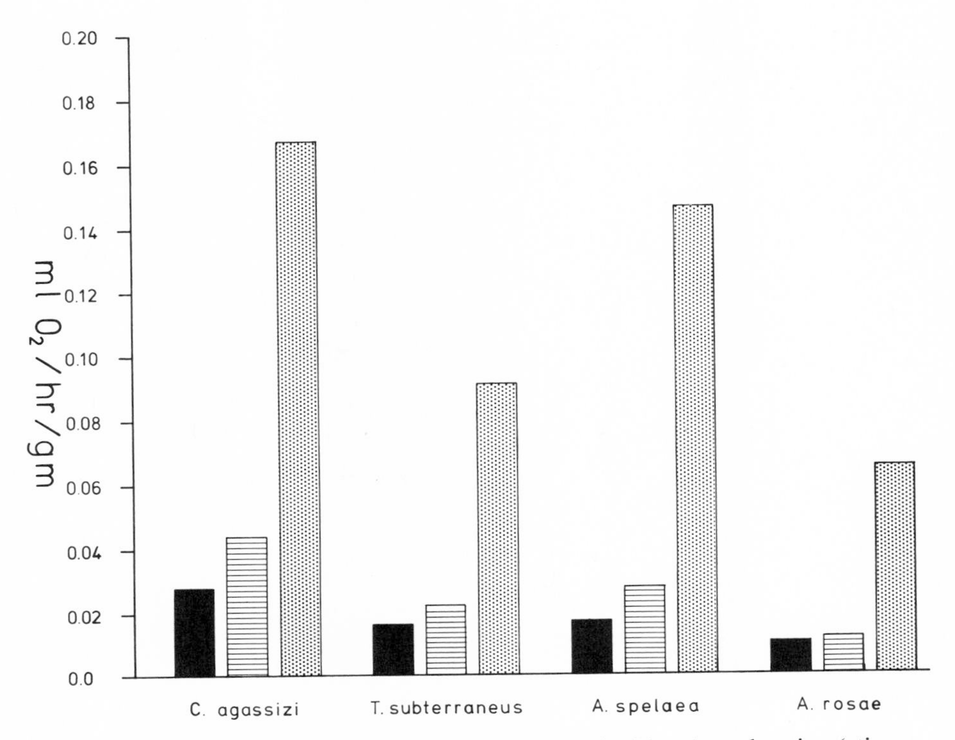

Figure 2–11 Standard (solid bars), routine (hatched bars), and active (stippled bars) metabolic rates in ml O_2/hr/gm for *Chologaster agassizi, Typhlichthys subterraneus, Amblyopsis spelaea,* and *A. rosae*. SMR is the metabolic rate of an inactive fish, RMR is that of a normally active fish, and AMR is that of a maximally active fish. (Data from Poulson 1963.)

few studies of metabolism of terrestrial cave animals parallels Dickson and Franz's findings. Kane and Poulson (1976) found that when starved, the cave beetle *Neaphaenops tellkampfi* showed about half the percentage of daily weight loss on sand and mud substrates than the cave beetle *Pseudanophthalmus menetriesii*. On litter substrates *N. tellkampfi* showed a nonsignificant reduction in weight loss compared to *P. menetriesii*. The most reasonable interpretation of these differences is that *P. menetriesii* is less food limited, its Collembola prey being locally abundant but patchily distributed. Other cave species that show reduced metabolic rates include the isopod *Caecosphaeroma serrata* and the amphipod *Niphargus longicaudatus* (Derouet 1959). Poulson (1964) reviews evidence for some other cases.

Reduction in metabolic rate is not universal in cave organisms.

Table 2–6 Standard Metabolic Rates (SMR, no activity) and Routine Metabolic Rates (RMR, normal activity) for cave and spring amphipods from Virginia and West Virginia. Rates are given in $\mu l\ O_2/0.01$ g/hr. (Data from Culver and Poulson 1971.)

Species	Habitat and locality	SMR	RMR
Gammarus minus	Organ Cave, W.Va.	3.3	3.4
	Benedicts Cave, W.Va.	4.1	4.3
	Coffmans Cave, W.Va.	2.5	2.6
Gammarus minus	Davis Spring, W.Va.	5.0	5.2
	U.S. 219 Spring, W.Va.	1.5	2.1
Stygobromus emarginatus	Court Street Cave, W.Va.	0.8	—
Stygobromus spinatus	McClung's Cave, W.Va.	2.9	—
Stygobromus tenuis potamacus	Seep, Fairfax Co., Va.	2.1	—

Schlagel and Breder (1947) found that the Mexican cave fish *Astyanax mexicanus* had nearly twice the routine metabolic rate (RMR) of surface populations of the same species. Caves where *A. mexicanus* are found have high organic input (Breder 1942), and the populations are not likely to be food limited. In a similar study Culver and Poulson (1971) found no evidence that cave populations of the amphipods *Gammarus minus, Stygobromus emarginatus,* and *S. spinatus* had lower metabolic rates than spring and seep populations of *G. minus* and *S. tenuis potamacus* (Table 2–6). Once again, available evidence suggests that these populations are not food limited (Culver 1971a).

Recent work by the French biologists Gibert and Mathieu raises questions about whether metabolic rates of many aquatic cave invertebrates are really lower than those of their nearest noncave ancestors. Some aquatic cave invertebrates arose from interstitial[1] populations rather than from epigean populations. The large amphipod genera *Niphargus,* primarily in Europe, and *Stygobromus,* primarily in North America, occur mostly in noncave habitats, with cave species apparently derived from these (Holsinger 1978). A similar situation may obtain with isopods (Magniez 1976) and flatworms. Gibert and Mathieu (1980) found that starved cave *Niphargus virei* depleted lipid and carbohydrate reserves more rapidly than interstitial populations of *N. rhenorhodanensis,* which suggests that *N. virei* has a higher metabolic

1. The term "interstitial" and its many partial synonyms are used in several ways. Here the term means any nonkarstic, noncave subsurface habitat, for example, riverine gravels. See chapter 7 for more details.

rate. Mathieu (1980) also found that cave populations of *N. rhenorhodanensis* were more active than interstitial populations, which also suggests a higher routine metabolic rate for cave populations. The most reasonable explanation for these data are that many interstitial habitats are even more food poor than cave habitats. Since the metabolic rate of *N. virei,* a cave species, is lower than that of any other amphipod reported in Wolvekamp and Waterman's (1960) review, it is likely that for *Niphargus,* both cave and interstitial environments are usually food poor.

Adaptation to High Moisture

Terrestrial cave species must adapt to a water-saturated atmosphere. Howarth (1980) reviews the evidence that there is an upper limit to humidity tolerance for surface-dwelling terrestrial arthropods and suggests that cave-limited species have evolved effective water excretory mechanisms that conserve salts and that this involves cuticular reduction, resulting in increased cuticular permeability. If cuticular permeability is increased as a result of adaptation to saturation or near-saturation—and there is no direct evidence on this point,—then in less humid conditions these species will lose water rapidly. This may explain the extreme moisture sensitivity of terrestrial troglobites (see Howarth 1980). Alternatively, cuticle reduction may be the result of selection for metabolic economy, since cuticle formation and maintenance incur a significant metabolic cost (T. L. Poulson, personal communication). Or cuticle reduction may be the result of relaxed selection. In any case, one would expect the evolution of mechanisms to insure that terrestrial cave organisms remain in areas of high humidity and low saturation deficit.

One clear case of evolution of humidity-detecting sensory structures is the highly developed internal antennal vesicles found in cave beetles in the family Catopidae (Peck 1977, Accordi and Sbordoni 1978). Peck found that compared to those of epigean species, the sensory hairs on the surface of antennal segments were longer in cave species of *Ptomaphagus* and that there was an increase in structural complexity of the internal antennal vesicle in the seventh antennal segment (Fig. 2–12). European cave catopids also show a structural elaboration of the internal antennal vesicles, although they are structurally quite different from those of North American species. From the experiments of Lucarelli and Sbordoni (1978), it is clear that a major function of these organs is humidity detection.

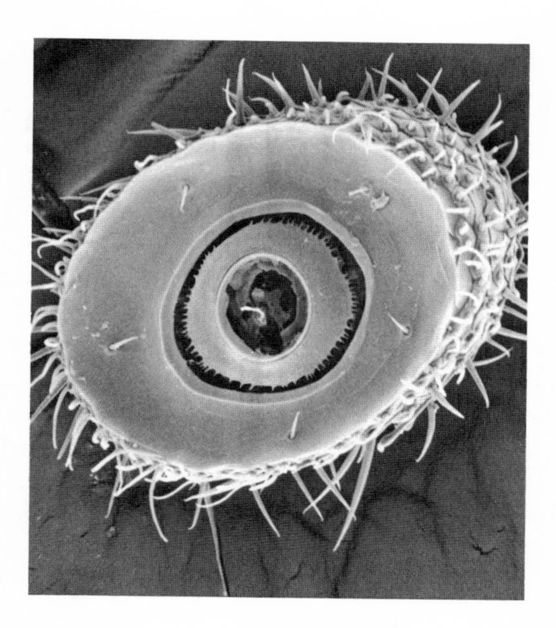

Figure 2–12 Scanning electron micrographs of internal antennal vesicle of *Ptomaphagus hirtus*. *Top,* antennal joint socket is in the middle, surrounded by ringlike slit opening (periarticular gutter) into sensory vesicles. *Bottom,* sensory pegs line the sensory vesicle. (Photos courtesy of Dr. Stewart B. Peck, Department of Biology, Carleton University, Ottawa.)

The importance of adaptation to high moisture, especially standing water, can be seen in cave Collembola in the genera *Sinella* and *Pseudosinella,* whose morphologies are probably the most thoroughly studied of any cave organisms. Christiansen (1961) compared Japanese and Nearctic *Sinella* and European and Nearctic *Pseudosinella,* and divided morphological characters into two groups—those that were cave independent and those that were cave dependent. Cave-independent characters, such as head chaetotaxy and labial papillae, showed no consistent changes between troglophilic (facultative) and troglobitic (obligate) cave species, but were useful in determining taxonomic and phyletic position. On the other hand, cave-dependent characters, such as antennal length, showed consistent differences between troglophilic and troglobitic species. The cave-dependent characters of highly modified troglobites are similar regardless of genus or geographic location. Cave-dependent characters that are both well studied and widespread are shown in Fig. 2–13; a larger list is given in Christiansen and Culver (1968). The mucro is at the end of the furcula, a tail-like structure folded under the abdomen that allows Collembola to "jump." This structure and the antennae are long, not only in cave forms but also in species that live on free surfaces. Christiansen suggests that these changes increase the animal's ability to detect predators and to escape quickly. Collembola are major prey items for many cave beetles (McKinney 1975, Kane and Poulson 1976) and probably for spiders and harvestmen as well.

Both highly modified cave species and deep soil species, such as the Onychiuridae, have an enlarged third antennal segment organ. Christiansen (1961) suggests that this organ is sensitive to humidity or temperature and notes that highly modified cave forms are very sensitive to both.

The tenet hair, empodial appendage, and unguis are part of the claw complex. Changes in this complex are clearly related to the ability of Collembola to walk on wet, smooth surfaces, wet clay, and water (Christiansen 1965). In many caves, pools act as Collembola traps and are likely to be a major cause of mortality. In an extensive series of observations Christiansen showed that the position of the claw on these surfaces changes with increasing adaptation. Of the animals with a stage 3 (highly cave-adapted) claw complex (see Fig. 2–13), 90 percent were able to move over a water surface, compared with only 4 percent of those in stage 1 (unmodified). Figure 2–14 shows the claw positions of non-cave-modified and cave-modified forms on wet clay. Similar differences between species in claw position on wet smooth surfaces and on water indicate the adaptive nature of the morphological changes.

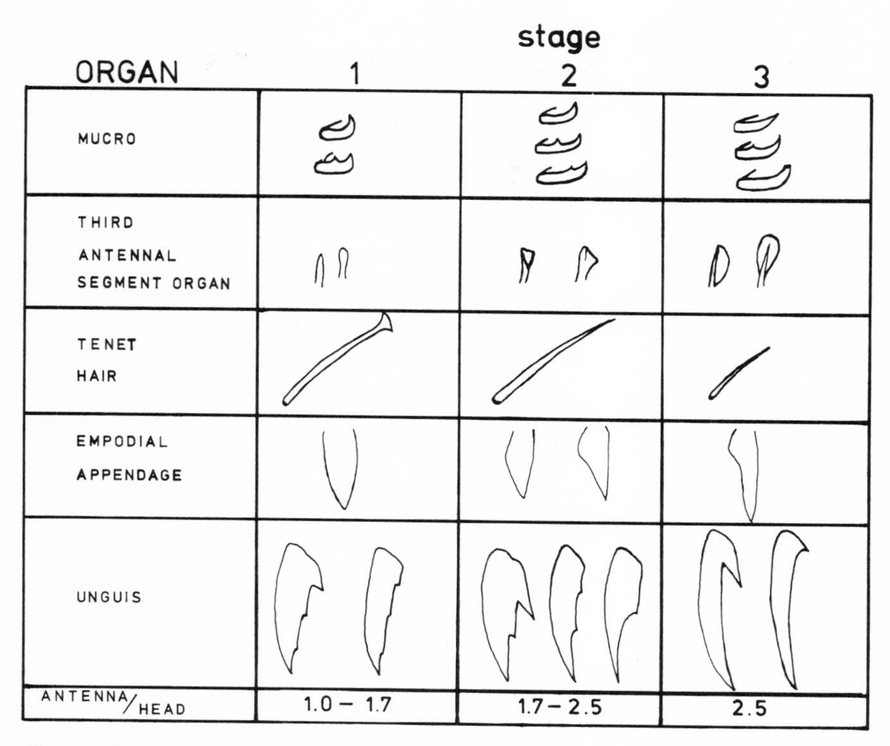

Figure 2–13 Modifications of Collembola morphology with increasing cave adaptation. Stage 1 represents species unmodified to cave life, stage 2 is intermediate, and stage 3 represents highly modified cave forms. (From Christiansen 1961.)

Finally, Christiansen (1961) provides some indirect evidence that the morphological changes increase fitness. Stage 1 species (with unmodified claw complexes) were found in less than ten caves, most within a range of less than 1,600 km^2, although their range in epigean habitats was much greater. Stage 2 species (intermediate modification) were found in more caves and over a greater area. However, they were either limited to caves or found in restricted epigean habitats, both cases reducing their dispersal potential. Stage 3 species (highly modified claw complexes) showed a broad variation in geographic range but were found in as many caves as stage 2 species. Speciation, local adaptation, and restricted movement account for differences in distribution of stage 2 and stage 3 species. But both are more successful in caves, at least as judged by geographic distribution, than stage 1 species.

Although cave Collembola may be modified to cope with food-poor environments, Christiansen (1965) found no evidence to support this.

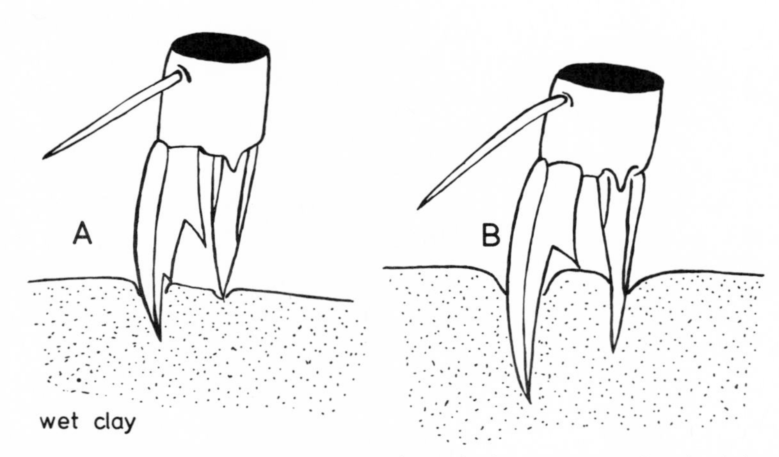

Figure 2–14 Positions and penetration of claw on wet clay in (*A*) non-cave-modified and in (*B*) cave-modified entomobryid Collembola. (From Christiansen 1965.)

Microfungi, which can be at least locally common (see Table 2–2), are probably their main food source.

Neoteny

Paedomorphosis, the retention of juvenile characters in the adult, has been reported for a variety of cave organisms. For example, *Speoplatyrhinus poulsoni,* the most cave-modified amblyopsid fish, shares some characters, such as body size and head size, with immature *Typhlichthys subterraneus*. The best-studied cases are the cave-dwelling plethodontid salamanders. Although there is no direct evidence, it is generally agreed that cave plethodontids are neotenic (retarded somatic development) rather than paedogenic (accelerated sexual development) (Gould 1977, Bruce 1979). Of the nine species of cave-limited salamanders, all but *Typhlotriton spelaeus* and *Gyrinophilus subterraneus* are neotenic and retain larval gills throughout their life (Brandon 1971, Besharse and Holsinger 1977). Of the other seven, only *G. palleucus* can be readily transformed by thyroxin treatments (Brandon 1971). The question that needs to be considered here is whether neoteny is an adaptation to low food supply. Wilbur and Collins (1973) suggest that it evolves in response to an unstable or harsh terrestrial environment compared to the aquatic environment. It is more likely that neoteny in cave salamanders, except for *Gyrino-*

philus, happens because of food scarcity in the terrestrial cave environment.

For *Gyrinophilus* the situation is more complex (Bruce 1979). Epigean stream-dwelling *G. porphyriticus* in the southeastern United States undergo a dietary shift but not a habitat shift at metamorphosis, becoming a specialist feeder on other species of salamanders. Bruce suggests that neoteny was selected for in the closely related troglobitic *G. palleucus* because of the absence of other salamanders that are major diet items of adult epigean *G. porphyriticus,* rather than because of food scarcity in the terrestrial environment. But some cave populations of *G. porphyriticus* do undergo a habitat shift at metamorphosis, in a manner consistent with the resource hypothesis. For example, in the large population of *G. porphyriticus* in Cope Cave in Lee County, Virginia, adults are almost exclusively terrestrial, feeding on an abundant oligochaete fauna on mud banks. One would not expect selection for neoteny if the resource hypothesis is correct, and there is no evidence for neoteny in this population. Neoteny clearly deserves more study, especially in *Gyrinophilus,* where there is considerable variation. But available evidence, largely anecdotal, is in agreement with the hypothesis that food scarcity is the main selective factor for neoteny.

Conclusions

Although the evidence is incomplete, several conclusions can be reached. First, many cave populations are adapted to scarce food supplies. Poulson's work on the Amblyopsidae is the most thorough and unassailable. Other groups, for example, crayfish (Cooper 1969), show adaptations to scarce resources in morphology and food-finding behavior. The high feeding efficiency of cave beetles that eat cricket eggs (Mitchell 1968, Kane and Poulson 1976) is most easily explained by strong selection for food-finding ability. The frequent finding that cave organisms are resistant to starvation also suggests that food scarcity is a dominant selective force.

But the second conclusion is that food scarcity is not a universally dominant selective force. Christiansen's elegant studies of entomobryid Collembola, which are at least numerically dominant in many terrestrial caves, strongly indicate that moisture and humidity, rather than food, have been the dominant selective factors. Other groups, such as *Stygobromus* amphipods, show little evidence of adaptation to low food supplies. Kane and Poulson's (1976) and Dickson and Franz's

(1980) demonstrations that metabolic rate and its correlates are related to levels of food supply indicate that there are at least differences in the intensity of selection for reduced metabolic rate among obligate cave species.

It is also likely that many troglophiles are not "troglobites in training." Many troglophiles have no known surface populations and are classified as troglophiles only because they show little sign of regressive evolution. Other troglophiles have no surface populations near cave populations. For example, *Gyrinophilus porphyriticus* is common in caves in the upper Powell Valley in Virginia and Tennessee, but no surface populations are known from this area. It is very rare to find a population of any cave organism that extends from the surface directly into a cave. In most cases it is unlikely that gene flow is retarding adaptation. It is likely that what are called troglophiles have been in caves for a shorter period of time than what we call troglobites, if only because troglophiles show less regressive evolution (but see chapter 4). Nonetheless, by any measure available, many troglophiles are very successful in caves. Many such as helomyzid and sciarid flies, are found in a large percentage of caves in a given area, and with large populations. More work is needed on these seemingly "uninteresting" species.

Finally, there is considerable evidence that many cave populations are not at an evolutionary equilibrium. If equilibria were widespread, then there would be no reason to expect differences in metabolic rate among amblyopsids based on their length of time of isolation in caves. There would be no reason to expect a fraction of a population of the crayfish *Orconectes pellucidus* to retain circadian rhythms (Jegla and Poulson 1968) or to expect that larvae of *Gyrinophilus porphyriticus* in caves would still occasionally show lunging behavior during feeding, even though it greatly reduces feeding efficiency (Culver 1975). The lack of equilibrium in turn implies either that selection has been weak or that selectable variation in these traits is scarce. It is impossible to know which is more important, but we do know that in many cases selection has had a long time to act; most terrestrial species in north temperate zones probably invaded caves during interglacials in the Pleistocene (Barr 1968), and aquatic species may be much older (Holsinger 1978).

3 Life History Tactics

Nearly all cave organisms for which any data are available show some or all of the following characteristics: delayed reproduction, increased longevity, smaller total number of eggs produced, and larger eggs—all features that are commonly associated with what is called K-selection. Since the demonstration by MacArthur (1962) of an analog to Fisher's Fundamental Theorem of Natural Selection in which population size can be maximized by natural selection, models of life history evolution and critiques of such models have been in full flood. Stearns (1977) has pointed out that the models have neglected critical aspects of the evolutionary process and that available empirical data are not extensive enough to distinguish among the various competing models. Because of the delayed maturity and low reproductive rates of cave organisms, it has not usually been possible to accumulate the sort of detailed information on life history parameters that is available for rapidly reproducing species such as *Drosophila*. Life histories of cave organisms should be of interest to modelers because they show the extremes of delayed reproduction and low reproductive rate. On the other hand, even in their current state of controversy, models of life history evolution should be of interest to cave biologists because they provide alternative explanations for the life history characteristics observed.

Two themes run through this chapter. The first is whether the observed life history parameters are adaptive. Consider the following data of Anderson and Watanabe (in Charlesworth 1980) for AR/AR karyotype female *Drosophila pseudoobscura:* the average number of eggs produced by a female was 83.1 under optimal conditions and 2.2 under yeast starvation. The stress imposed by severe food limitation would seem almost inevitably to cause a slowing of growth, a reduction in reproduction, or increased mortality. This is a result of the constraints imposed by lack of available energy and can be called adaptive only in the most trivial sense. This is not to say that cave organisms are slowly starving to death and that their life history traits are not adaptive. Quite the contrary; but a healthy skepticism about universal adaptiveness of life history traits throws those traits that are truly adaptive into bolder relief. The clearest example of this is the large egg size of cave organisms, which is almost certainly adaptive, because larger eggs result in larger offspring that are more resistant to starvation and perhaps to predation.

The second theme of this chapter is that delayed reproduction can be explained by several models, most of which have been recently summarized by Charlesworth (1980). There is no simple correspondence between density dependence of populations and the prediction of delayed reproduction, on the one hand, and density independence of populations and the prediction of accelerated reproduction, on the other. Charlesworth points out that not all forms of density dependence lead to selection for delayed reproduction (or for increasing reproductive effort with age). In particular, a density-dependent mortality factor applied equally to each age class does not result in selection for delayed reproduction. Among the factors that do lead to such selection for delayed reproduction are: (1) density-dependent juvenile survival or fecundity (Charlesworth 1980); (2) density-dependent adult fecundity (Charlesworth 1980); (3) fluctuating juvenile survival (Schaffer 1974a); and (4) long periods in which r, the intrinsic rate of increase, is negative (Mertz 1971). Low total reproductive effort, another characteristic often associated with K-selection, may result when juvenile mortality is high (Schaffer 1974b) or when adult mortality is low and r is low (Goodman 1974).

The life history models considered in this chapter are limited to those with a clearly defined selective basis. That is, it is assumed that "life histories evolve as a result of gene frequency changes within populations, under the control of natural selection" (Charlesworth 1980, p. 205). Some optimization models of life histories require group selection, and the requirements of others are not clear. I adhere generally to Charlesworth's viewpoint in what follows.

Although there is considerable information on egg numbers and size distributions in the cave biology literature, there are few studies that are detailed enough to allow estimation of life history characteristics. In the rest of this chapter I will consider four of the most complete comparative studies, the first two of which are laboratory studies. By far the most complete laboratory study of terrestrial cave populations is Deleurance-Glaçon's (1963) work on European leiodid cave beetles, supplemented by Peck's (1975a) laboratory study of leiodid populations in North American caves. Rouch's (1968) laboratory studies of aquatic European harpacticoid copepods will then be reviewed. The two most complete field studies are Poulson's (1963, 1969) work on North American amblyopsid fish, and Cooper's (1975) study of North American crayfish.

Leiodid Beetles

Adult cave leiodid beetles are scavengers and saprophages. The most effective baits for them are rotten meat and dung (Peck 1973a, 1975b), and in Mammoth Cave considerable numbers can be found on dung. Although food is generally scarce for the beetles, those species that can utilize mammalian dung at least occasionally have abundant resources. The adults probably have few predators in most caves, occasionally being taken by salamanders (Peck and Richardson 1976), staphylinid beetles, and possibly the larvae of fungus gnats (Peck and Russell 1976). The free-living, active stages of the beetle larvae are carnivorous, but their prey are unknown (Deleurance-Glaçon 1963). Being soft-bodied, the larvae and pupae are more susceptible to predation. Peck (1973a) reports predation by Collembola and mites.

Mortality and fecundity data are available for four species, summarized in Table 3–1. Considering only the three European Bathysciini for the moment, there are large differences in fecundity among the species. For *Isereus colasi* and *I. serullazi,* in line with what we expect for a species with low food availability, egg production is very low. Deleurance-Glaçon states that these species have large eggs, but she gives no measurements. On the other hand, *Speonomus delarouzeei* produces many small eggs, ten times as many as *Isereus*. Direct comparisons with the North American *Ptomaphagus hirtus* are difficult to interpret because of its different phyletic history and different laboratory rearing conditions, but it is intermediate in egg output between the two European genera.

There are two possible interpretations of the fecundity differences. First, it could be argued that *Isereus* has been in caves longer than

Table 3–1 Life history characteristics of cave leiodid beetle species for which both mortality and fecundity data are known. Rearing temperature of European species (*Speonomus delarouzeei, Isereus serullazi* and *Isereus colasi*) was 9°C; rearing temperature of American species, *Ptomaphagus hirtus*, was 12.5°C. Assuming a Q_{10} of 2, and multiplying the figures for the European species by a factor of 0.825, the species can be compared at equivalent temperatures. (Data modified from Deleurance-Glaçon 1963 and Peck 1975a.)

	Eggs		Larvae			Pupae		Adults	
Species	No. eggs/ female/yr	Time to hatching (days)	No. instars	Larval duration (days)	% of time feeding	Pupal duration (days)	Pupal duration/ larval duration	Female mortality/yr	Male mortality/yr
Ptomaphagus hirtus	51	18.5	3	42	70	32	0.76	0.42	0.25
Speonomus delarouzeei	131	51	2	167	18	50	0.29	0.33	0.23
Isereus serullazi	13.8	117	1	99	0	116	1.17	—	0.29
Isereus colasi	11.8	117	1	94	0	117	1.24	0.35	—

Speonomus and that *Speonomus* populations are therefore less adapted and less fit. Although this view may be correct, there is no unambiguous evidence to support it. A second hypothesis is that the changes in larval morphology, the suppression of larval feeding period, and the more "regressive" characters of *Isereus* may all be the result of a single selective regime, and that species such as *Speonomus* may be under a different selective regime. *Speonomus delarouzeei* (and *P. hirtus*) may face a lesser food shortage and may have retained the ability to increase rapidly when a rich food source such as mammalian dung is available. Under Mertz's (1971) scheme of classifying species by whether their intrinsic rate of increase is usually positive or usually negative, it may be that r is usually negative for *Isereus* and usually positive for *Speonomus* and *Ptomaphagus*. Deleurance-Glaçon mentions in passing that there are considerable ecological differences among the species, but she gives no particulars. The second hypothesis has no more direct support than the first, but it is at least possible that the life history characteristics of *S. delarouzeei* are adaptive rather than being an inferior version of the *Isereus* life history characteristics. All species listed in Table 3–1, including *S. delarouzeei,* are troglobites (Laneyrie 1967), so there is no possibility of "contamination" by genes from surface populations.

The most apparent pattern in adult mortality is that males consistently have lower rates than females. Time to adult emergence is much shorter for *P. hirtus,* but this is due at least in part to higher rearing temperatures and different food. The time to adult emergence for *S. delarouzeei* (268 days) is less than that of *I. serullazi* (332 days) and *I. colasi* (328 days), which is in line with their fecundity differences.

The most fascinating and unexpected aspect of the *Isereus* life history is the suppression of larval feeding. While a variety of models can explain delayed reproduction (see the beginning of this chapter), suppression of feeding would seem to increase larval mortality, and I know of no model that predicts selection for increased larval mortality. Either suppression of feeding is maladaptive, which is unlikely in view of the extensive morphological changes of the larvae that accompany suppression, or we must look for some explanation that involves selection. I think two factors have led to suppression of larval feeding. First, live prey available to the larvae are probably scarce, so the energy gained by feeding is probably not great. Second, the increase in the pupal period in *Isereus* suggests that the pupal cell protects the organism from predation. For suppression of feeding to be favored, the gain in survivorship from predator escape should exceed the loss from lack of food. As an example of the potentially high mortality rates of immature stages, consider the following life table based on *I. colasi* with an adult longevity of four years of reproduction after the first year of adult life and with constant adult mortality; b_x, represents the number of female eggs produced at age x, and l_x, represents survivorship to age x.

Age	Adult age	l_x	b_x
0	—	1	0
1	0	l_i	0
2	1	$(0.65)l_i$	5.9
3	2	$(0.42)l_i$	5.9
4	3	$(0.27)l_i$	5.9
5	4	$(0.18)l_i$	5.9

If the population size is stable and r is zero, the net reproductive rate

$$R_o = \sum_x l_x b_x \qquad (3\text{-}1)$$

is 1. For the above life table, $R_o = 1$ when survival to the adult stage, l_i, is 0.11. This at least indicates that high mortality of immature stages is possible.

Harpacticoid Copepods

Harpacticoid copepods are common in the sediments and gravels of many cave streams and pools. As Delamare-Deboutteville points out (quoted by Rouch 1968, p. 13), harpacticoids are "adapted to creep on the substrate or to saunter rapidly among the interstices of the sediments" (my translation). They are common in a wide variety of subsurface habitats besides caves, including sands bordering fresh and salt water. Compared to cave organisms at higher trophic levels, harpacticoids probably suffer high mortality. Their soft-bodied, wormlike morphology is likely to make them susceptible to the vagaries of currents and to the mechanical abrasion of sediments. In addition, many populations suffer predation from groups such as amphipods, isopods, and crayfish. The food of harpacticoids, organic debris and microorganisms, is generally scarcer in caves than in epigean habitats (Poulson 1964, Gittleson and Hoover 1970).

Most of what is known about cave harpacticoid life cycles comes from Rouch's (1968) study of French species. Rouch provides no mortality data, but he does provide detailed information on fecundity of both subsurface and epigean species. In Figure 3–1, egg numbers and egg size are plotted against body size for five cave species, four high-altitude species in streams and ponds, and four ubiquitous low-elevation species in streams and lakes. One of the low-elevation species, *Bryocamptus zschokkei,* is also found in caves, but Rouch reports data only for epigean populations. Besides the ecological classification, Rouch distinguishes three phyletic lines that include both cave and epigean species: *Nitrocella* and *Nitocra* in the Ameridae; *Bryocamptus* in the Canthocamptidae; and *Elaphoidella* in the Canthocamptidae. In a food-poor environment such as caves, one would expect reduced number of eggs per brood and increased egg size. These predictions are in general borne out (Fig. 3–1*A*). All cave species have less than ten eggs per brood, and all low-elevation epigean species have more than ten. This is due in part to the smaller size of the cave species, but in the *Elaphoidella* line, cave species of the same size as epigean species have fewer eggs. Surprisingly, high-altitude species have even fewer eggs, perhaps because of food scarcity in this environment as well. Egg diameters show a similar pattern (Fig. 3–1*B*). The eggs of high-altitude and cave species have larger diameters than those of epigean species, except for the epigean *Canthocamptus staphylinus,* which has a much larger body. There is little doubt that the differences shown by cave species in egg volume and number are the result of adaptation to a food-poor environment.

Although mortality rates are probably density dependent, because of predation, this density dependence does not necessarily lead to delayed reproduction. However, low food supplies may result in density-dependent fecundity, which should lead to delayed reproduction (Charlesworth 1980). Rouch provides data on postembryonic development and adult longevity for one epigean species, *Bryocamptus zschokkei,* and two cave species, *B. pyrenaicus* and *Nitrocella subterranea*. Postembryonic development is 4 weeks for *B. zschokkei,* 10–13 weeks for *B. pyrenaicus,* and 11–16 weeks for *N. subterranea*. Adult longevity is 8.7 months for *B. zschokkei*, 19.6 months for *B. pyrenaicus,* and 14.6 months for *N. subterranea*. While the differences between *Bryocamptus* and *Nitrocella* may be the result of phyletic differences, the differences between *B. zschokkei* and *B. pyrenaicus* are probably caused by selection for delayed reproduction in *B. pyrenaicus*.

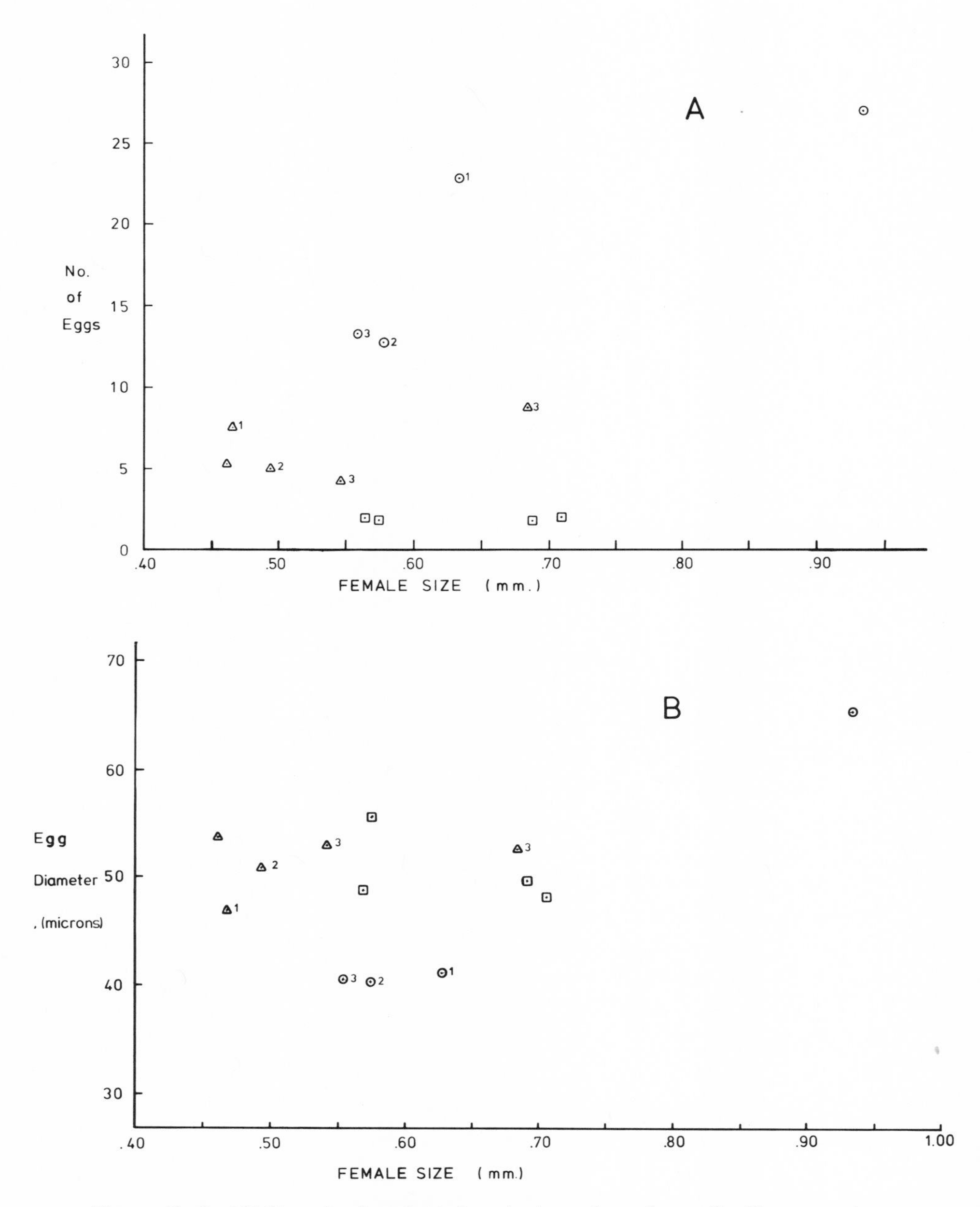

Figure 3–1 (*A*) Female size plotted against number of eggs for European harpacticoids. (*B*) Female size plotted against egg diameter. Squares are species in high-altitude streams and ponds, triangles are cave species, and circles are low-altitude epigean species. The three phyletic lines labeled are: 1, *Nitrocella* and *Nitocra;* 2, *Bryocamptus,* and 3, *Elaphoidella.* (Data from Rouch 1968.)

Rouch's data on egg number and egg diameter can be used to estimate reproductive effort. Compared to other cave organisms, in these species adult mortality and r are high, so reduced reproductive effort is less likely (Goodman 1974). As indicated in Figure 3–2, there is no difference between cave and low-altitude epigean species in reproductive effort per brood except for the epigean *Canthocamptus staphylinus,* whose reproductive effort is nearly an order of magnitude higher than that of any other species. High-altitude epigean species show some indication of reduced reproductive effort. If the number of broods is taken into account, cave species have lower reproductive effort, but the females are smaller as well (Table 3–2). Given the small number of species, there is no conclusive evidence for reduced lifetime reproductive effort.

Amblyopsid Fish

In addition to the modifications in morphology and physiology discussed in chapter 2, Poulson (1963, 1969) has provided considerable in-

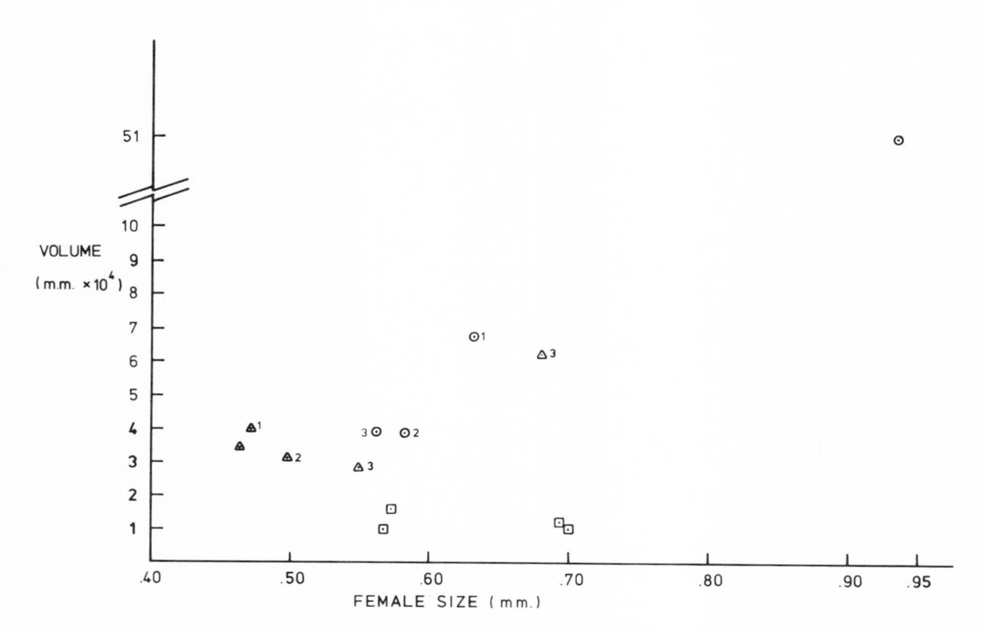

Figure 3–2 Reproductive effort per brood, measured as total egg mass volume plotted against female size. Labels as in Fig. 3–1. (Data modified from Rouch 1968.)

Table 3–2 Lifetime reproductive effort for epigean *Bryocamptus zschokkei* and cave *B. pyrenaicus* and *Nitrocella subterranea*. (From Rouch 1968.)

Species	Habitat	Mean length of female (mm)	Egg vol./brood (mm^3)	No. broods	Total egg vol. (mm^3)
Bryocamptus zschokkei	Epigean	0.63	6.9×10^{-4}	14	9.7×10^{-3}
Bryocamptus pyrenaicus	Cave	0.47	4×10^{-4}	16.7	6.7×10^{-3}
Nitrocella subterranea	Cave	0.50	3×10^{-4}	9.2	1.2×10^{-3}

formation on life cycle modifications of cave amblyopsid fish. Fecundity and longevity data are summarized in Figure 3–3 and Table 3–3, in which the species are listed by increasing evolutionary time in caves. *Amblyopsis* is also a separate phyletic line, albeit closely related, and has been in caves the longest time. *Chologaster cornuta,* found in swamps and small streams in the Coastal Plain, can be considered the base line for comparing varying levels of cave adaptation. *C. agassizi* is typically found in springs as well as occasionally in caves. The other three species are found only in caves. All species show to varying degrees an annual peak in reproduction that corresponds to the time of high food input (Poulson 1963).

Typhlichthys subterraneus and both *Amblyopsis* species show delayed maturity compared to the *Chologaster* species (Fig. 3–3), with *Typhlichthys* maturing after two years, and *Amblyopsis* after three. The most likely selective factor favoring delayed maturity is density-dependent adult fecundity (see below), but long periods of population decline should result in selection for delayed reproduction as well (Mertz 1971). The mean number of ova is smaller for cave species, and ovum volume is larger (Table 3–3), which corresponds to the classic pattern of a K-selected species. The oldest species, *A. rosae,* has the smallest number of ova, but ovum volume is somewhat smaller than for other cave species. *Chologaster agassizi* does not fit the pattern, having more ova than *C. cornuta.* However, very little is known about the ecology of *C. cornuta,* and population growth may be density-dependent for it as well. It is associated with a large number of potential predators and competitors (Poulson 1963).

Reproductive life span presents a more complicated pattern. The major difference in life cycle between *C. cornuta,* the swamp fish, and

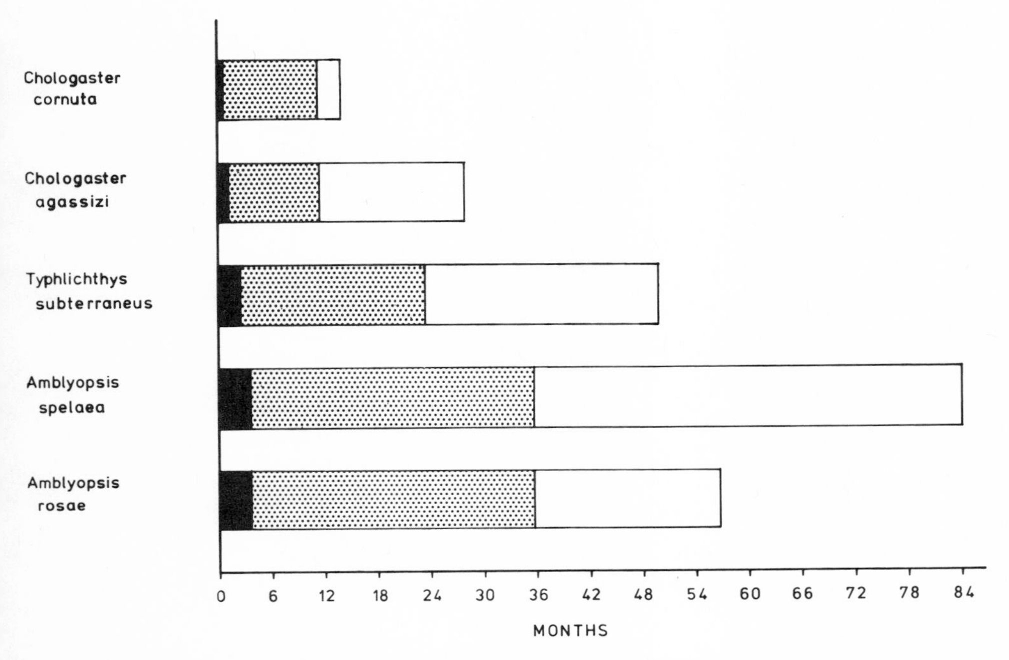

Figure 3–3 Life spans of species of amblyopsid fish. Black bar is time to hatching, speckled bar is time from hatching to first reproduction, open bar is reproductive life span. (From Poulson 1963.)

C. agassizi is that *C. cornuta* is always semelparous and *C. agassizi* is sometimes iteroparous. The conditions that favor iteroparity are well known. For a density-independent population with constant adult survival from one age class to another (P) and identical probabilities of survival to reproduction, iteroparity will be favored when age-specific fecundity, m_i, exceeds the value given by the equation (Charnov and Schaffer 1973):

$$m_i/m_s = (1 - Pe^{-r}) \tag{3-2}$$

where m_s is the fecundity of the semelparous population, and r is the intrinsic rate of increase. Iteroparity is favored when adult survival is high and r is low. For density-dependent populations, the equation reduces to (Charlesworth 1980):

$$m_i/m_s = (1 - P) \tag{3-3}$$

Table 3–3 Reproductive effort of amblyopsid fish. (Data modified from Poulson 1963.)

Species	Habitat	Mean no. ova per female	Ovum vol. (mm^3)	Female wt. (gm)	Average no. reproductions	Maximum no. reproductions	Reproductive effort/ gm of female (mm^3)		
							Per brood	Lifetime observed	Lifetime maximum
Chologaster cornuta	Swamp	98.0	0.61	0.93	1.0	1	64	64	64
Chologaster agassizi	Spring	152.1	2.80	2.87	2.0	2	148	297	297
Typhlichthys subterraneus	Cave	49.8	5.20	0.86	1.5	3	301	452	903
Amblyopsis spelaea	Cave	69.5	5.20	6.88	0.5	5	52	26	260
Amblyopsis rosae	Cave	23.0	4.50	1.25	0.6	3	83	50	249

and iteroparity is favored when adult survival is high. Since *C. agassizi* is in a more stable environment and has fewer potential predators than *C. cornuta*, iteroparity is expected. *Typhlichthys subterraneus*, *A. spelaea*, and *A. rosae* all show increases in reproductive life span, especially *A. spelaea*, which has a reproductive life span of five years (Fig. 3–3). Because reproduction occurs in more than one age class, the populations are iteroparous. However, it is unlikely, at least for the *Amblyopsis* species, that any individual reproduces more than once (Table 3–3). For example, Poulson (1963) found that only 10 percent of the reproductively mature *A. spelaea* actually reproduced in one year. Although some individuals may reproduce more than once, it seems more likely that most reproduce once at most. Regardless of whether repeated reproductions occur, a significant proportion of the individuals in the population never reproduce (Table 3–3), and the average number of reproductions per lifetime for *A. spelaea* is only 0.5. This indicates both that selection is potentially strong and that the environment is so food poor that many individuals never reproduce. It might be possible to invoke a group-selection or kin-selection argument to explain why failure to reproduce is adaptive, but this would depend on either shared rearing of young or high population extinction rates, for both of which there is no evidence. It seems most likely that many of the fish are simply slowly starving to death or are unable to obtain enough food to reproduce.

Table 3–3 gives estimates of reproductive effort. Whether one considers per-brood reproductive effort or maximum or observed lifetime reproductive effort, the same pattern emerges. In the *Chologaster-Typhlichthys* line, reproductive effort increases with increasing cave adaptation. On the other hand, *Amblyopsis* shows a reduced reproductive effort compared to both *T. subterraneus* and *C. agassizi*. Since reproductive effort should be low when adult mortality and r are low (Goodman 1974), why does *T. subterraneus* show a high reproductive effort? It is clear from the data in Table 3–3 and Figure 3–3 that the maximum potential rate of increase for *T. subterraneus* is higher than for either *A. spelaea* or *A. rosae*, but this simply changes the question to why r is higher for *T. subterraneus*. There are some habitat differences between the species. In particular, *T. subterraneus* can occur in smaller streams than *Amblyopsis*, and these smaller streams may have greater food supplies. However, even where *T. subterraneus* occurs in large, food-poor streams, it maintains a high reproductive effort (Poulson 1963). Based on available information, it would seem that the life cycle of *T. subterraneus* is not as adapted to cave conditions as that of *Amblyopsis*, presumably because *T. subterraneus* has been in caves a shorter time.

Poulson also provides evidence for density-dependent population growth. As shown in Figure 3–4, the age distributions of all of the cave species are skewed toward larger age classes. Assuming a stable age distribution, this is most likely to result from density-dependent factors. More direct evidence of density dependence comes from Poulson's (1969) study of *A. spelaea* in Upper Twin Cave, Indiana (Table 3–4). In a five-year study he found that total metabolic demand of the population varied less than either biomass or numbers of individuals. For food-limited populations, total metabolic demand should more accurately reflect carrying capacity than either numbers or biomass. A plot of the log of metabolic demand at time $t + 1$ against the log of metabolic demand at time t can be used to assess density dependence in a manner analogous to Tanner's (1966) use of population size. With density dependence, the slope should be less than 1, which it is for the *A. spelaea* population (slope = −0.66, S.D. = 0.54).

Finally, there is evidence that mean density and, by implication, carrying capacity, increases over evolutionary time as expected if *K*-selection is occurring. Poulson (1969) gives the following figures for

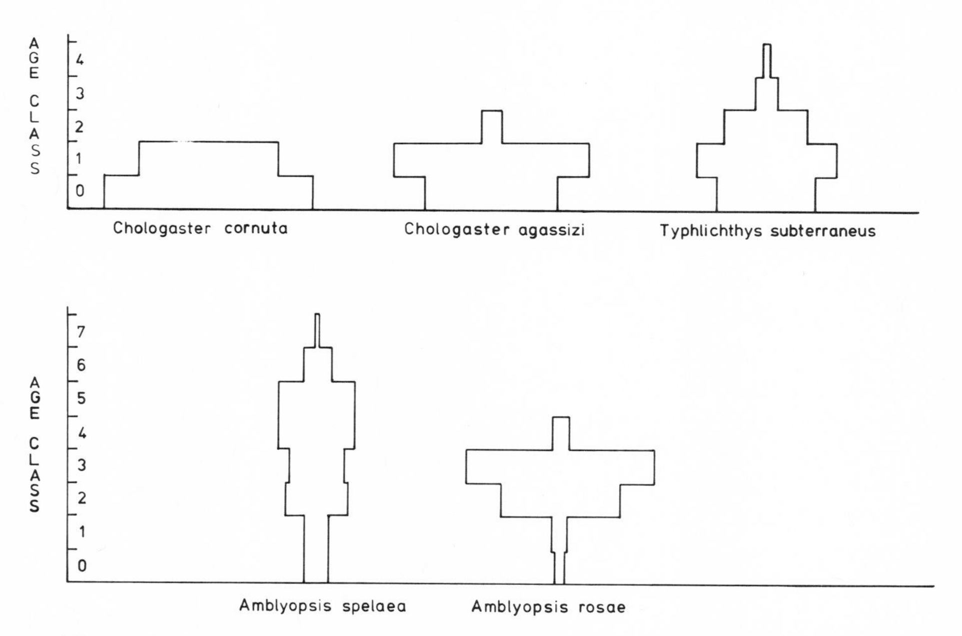

Figure 3–4 Age structure of five species of amblyopsid fish. (Modified from Poulson 1963.)

Table 3–4 Population characteristics of *Amblyopsis spelaea* in Upper Twin Cave, Indiana. The population was sampled once a year for five years. (Modified from Poulson 1969.)

	Total no. fish	Total weight (gm)	Total metabolic rate (cc O_2/day)
Mean	111	421	162
Coefficient of variation $\left(\frac{\sigma^2}{\bar{X}}\right)$	303	402	128
Range	84–130	363–455	149–180

number of fish per hectare in caves: *C. agassizi,* 50; *T. subterraneus,* 265; *A. spelaea,* 520; *A. rosae,* 1500.

Orconectes australis australis Crayfish

In the most extensive field study to date of the life cycles of cave organisms, Cooper (1975) analyzed crayfish populations in Shelta Cave, Alabama, over a period of 6 years. Because the *Orconectes australis australis* population in that cave represents the extreme in longevity and delayed reproduction, it is an appropriate final example.

Shelta Cave consists of several large, connected galleries with lakes. A very rich aquatic fauna is present, including three species of cave-limited crayfish (one *Orconectes* and two *Aviticambarus*), and two predators on young crayfish: the fish *Typhlichthys subterraneus* and the salamander *Gyrinophilus palleucus*. The most common crayfish was *O. australis australis,* with a population estimated at around 1,000 (Copper 1975). Over the course of the six years, Cooper marked over 900 *Orconectes* with codes unique to each individual, most of which were retained through molts. Because of a high recapture rate of marked individuals, Cooper was able to estimate growth rates of a rather large number. By summing up the times taken for each growth increment, one can calculate longevities (Table 3–5). Using the average rate of growth, Cooper found that the average time taken to reach the maximum carapace length of 47 mm was 176 years. Using the maximum growth rate observed for each size interval, the life span was 37 years. No life span can even be calculated by using the minimum growth rate, because some crayfish shrank in size! Cooper himself urges caution in interpreting these figures, but they are most remarkable, especially

Table 3–5 Growth rates and longevity in *Orconectes australis australis*. Rates are expressed in mm per month. The smallest size class (15.7 mm) is approximately that of newly hatched individuals. (From Cooper 1975.)

No. samples	Change in carapace length (mm)	Size increment (mm)	Max. rate	Months	Av. rate	Months
3	5.6 to 10.7	5.1	0.24	21.3	0.12	42.5
*	10.7 to 16.2	5.5	0.31	17.7	0.17	32.4
3	16.2 to 21.0	4.8	0.38	12.6	0.22	21.8
6	21.0 to 30.2	9.2	0.14	65.7	0.09	102.2
31	30.2 to 39.2	9.0	0.13	69.2	0.04	225.0
13	39.2 to 47.0	7.8	0.03	260.0	0.0005	1,695.7
Total months				446.5		2,119.6

* Rates determined by interpolation from adjoining size classes.

since the data are the most extensive available and the most carefully taken. The most reasonable interpretation of the growth rates given in the table is that the average growth rates in fact represent the actual growth rates of the population during the time it was sampled, and that the maximum growth rates approach growth rates under optimal conditions, at least for Shelta Cave.

Size at first reproduction can be estimated in several ways. Females are sexually mature when the carapace is 38–39 mm long, but the carapaces of the only successfully reproducing females were between 45 and 47 mm long. One smaller individual (carapace length 38 cm) had the remnants of an unsuccessful clutch. Translated into age, sexual maturity is reached in 35 years under average conditions, and in 16 years under optimal conditions. The age of reproduction occurs at 105 years for individuals with average growth rates and at 29 years for individuals with maximum growth rates.

During the entire study Cooper (1975) found only two reproducing females, one with 54 attached young and one with 78 attached young, which are low figures compared to most epigean species. However, he also found indirect evidence that reproducing females sequester themselves. The proportion of large females in the population was observed to drop during the winter. Assuming a constant sex ratio for the whole population, the sex ratio of the observed population can be used to estimate the fraction of large females reproducing. At a maximum, the fraction of large females reproducing is 0.67 and some of these reproductions are probably unsuccessful, because of egg resorption, for example.

Given delayed maturity and relatively small clutch sizes, the question is whether the population can maintain itself ($R_o \geq 1$) under the conditions Cooper observed. To answer this, one needs mortality estimates (see equation 3-1), which can be figured from the decline in frequency of marked individuals in Cooper's last sample, taken 20 months after the last marking. Assuming that a constant fraction of the population dies each month, irrespective of age, the monthly adult mortality rate is 0.020. This is an overestimate of mortality because some emigration may have occurred. More important, it is an underestimate because it neglects juvenile mortality, since few juveniles are marked. Juvenile mortality is higher than adult mortality because juveniles are preyed upon by *Typhlichthys subterraneus* and *Gyrinophilus palleucus*.

The estimate of mortality rate, with the caveat that it is too low, can be compared to mortality rates required to produce a nondeclining population for several idealized populations of *O. australis australis*. Four such populations were considered, differing in age of reproduction of females (Table 3–6). For simplicity, males were not considered. The ages of reproduction in the table correspond to the observed ages of reproduction *and* to reproductive maturity both for populations of indi-

Table 3–6 Maximum mortality to maintain a nondecreasing population of *Orconectes australis australis* in Shelta Cave, assuming constant mortality for four idealized populations differing in age of reproduction. Clutch size is assumed to be 40 female eggs; CL = carapace length.

Population growth conditions	Age of reproduction (yrs)	Maximum mortality		Observed mortality	
		Month	Year	Month	Year
Maximum, reproduce at 39.2 mm CL	15	.026	.218	.030	.333
Maximum, reproduce at 45 mm CL	29	.011	.119	.030	.333
Average, reproduce at 39.2 mm CL	35	.009	.100	.030	.333
Average, reproduce at 45 mm CL	105	.003	.034	.030	.333

viduals with average growth rates and for populations with maximum growth rates. All individuals are assumed to reproduce once and to produce forty female eggs. Individuals may reproduce more than once, but because considerable mortality probably accompanies reproduction, additional reproductions would be rare and would add relatively little to the growth rate. To compensate for this, clutch size is put somewhat higher than observed. Regardless of these details, a clear pattern emerges. In all cases, observed mortality is much higher than the mortality required to maintain population size. Therefore, the population in Shelta Cave is probably declining, albeit slowly, corresponding to Mertz's idealized population in which r is usually negative with occasional bursts of positive r. The delayed maturity is most likely the result of selection to slow the rate of decline in a population that is generally declining.

Summary

The life history patterns of fish and crayfish species, which are higher up in the food web and relatively free of predators, are easier to interpret than the patterns of lower species. Both fish and crayfish show signs of severe food limitation. Many individuals of the cave-limited species, with the exception of *Typhlichthys subterraneus*, fail to reproduce even once. At least for the crayfish studied by Cooper, there was evidence of failed reproduction and of egg resorption. But there was also strong evidence of an adaptive response to severe food limitation. Fewer, larger offspring were produced, and reproduction was delayed. The only incongruity was in the pattern of reproductive effort in the amblyopsid fish. With low adult mortality and low population growth rates, reproductive effort should be low. Two predictions can be made in this regard. First, the *Amblyopsis* line, which has apparently been in caves longer than the *Chologaster-Typhlichthys* line, should have a lower reproductive effort. Second, within the *Chologaster-Typhlichthys* line, reproductive effort should decline with increasing cave adaptation. The first prediction was confirmed by the data, but the second was not. Although there is considerable room for doubt, it is probably true that *Typhlichthys* is less adapted to the cave environment than the *Amblyopsis* species because *Typhlichthys* has had less time to adapt.

Some of the leiodid beetles, especially *Isereus serullazi* and *I. colasi*, show the classic pattern of delayed reproduction, with the unexpected twist of suppressed larval feeding. Although there is no comparable data available for surface-dwelling species, the life history patterns of *Ptomaphagus hirtus* and *Spenomus delarouzeei* are clearly less modified than those of *Isereus*. This may be caused by differences in the length of time these species have been in caves; it is just as likely that niche differences have allowed *Ptomaphagus* and *Speonomus* to exploit a more abundant food supply, such as mammalian feces. This is unlike the situation with fish and crayfish, where there is no obvious alternative food source that is at all abundant.

Compared to closely related surface species, all cave harpacticoids have fewer but larger eggs, indicating food scarcity. However, the differences are not extreme. Harpacticoid species from high-elevation surface habitats have even fewer and larger eggs than the cave species. On a per-brood basis, there is no indication of reduced reproductive effort for cave harpacticoids, which is consistent with the conjecture that their rate of increase is higher than that of many other cave groups and thus less likely to show reduced reproductive effort.

4 Regressive Evolution

To the nonspecialist, the hallmark of a cave organism is the absence of or great reduction in eyes and pigment. The mechanism by which this regressive evolution has occurred has been a subject of controversy from the time of discovery of the first blind cave animals up to the present. Regressive evolution is by no means unique to caves, but this fact is usually not mentioned in the cave biology literature. Prominent examples outside of caves include loss of flight in birds with the loss or reduction of tail or crista sterni (Kosswig 1948), limb loss in tetrapods (Lande 1978), and extreme reduction in eye size of some fish in turbid waters.

What makes regressive evolution in cave animals especially interesting is the very real possibility that selection may play a minor role compared to the accumulation of neutral mutations and genetic drift. The question of whether selection or neutral mutation plays the dominant role cannot be answered on the basis of our current knowledge, and the major theme of this chapter is that both explanations are plausible. With the exception of Wilkens (1971), there has been a recent bias toward selective explanations, involving selection for increased metabolic economy (Poulson 1964) or indirect effects of pleiotropy (Barr 1968). My bias here is toward documenting the plausibility of a neutral mutation explanation.

Whichever theory or theories is correct, it must be consistent with the known facts of the range and amount of regressive evolution and its underlying genetics. The chapter begins with a review of available data. Because of their importance to neutral mutation theory, three questions will be answered with particular care: (1) how long have organisms been in caves? (2) how many loci are involved? and (3) how large are cave populations?

The Extent of Regressive Evolution

The fauna of caves, whether facultative or obligate, is not a random sample of the epigean fauna. In any list of cave faunas one is struck by the preponderance of certain groups: for example, planarians, pericarid crustaceans, spiders, millipeds, Collembola, carabid beetles, and Diptera in West Virginia (Holsinger, Baroody, and Culver 1975). Much of this nonrandomness is due to the absence of herbivores, but it is also reasonable to expect that many ancestors of cave populations were "preadapted" in the sense that they did not rely primarily or exclusively on vision for locating food and for reproduction. For example, many Collembola are preadapted by virtue of living in soil or litter.

The best-known and often presumed typical example of regressive evolution is the Mexican cave characin fish, *Astyanax mexicanus*, which has relatively large-eyed ancestors (Fig. 4–1) with which it can interbreed. However, Collembola may be more typical of regressive evolution in caves. Collembola have a maximum of eight corneas per eye (Christiansen and Bellinger 1980–81), and the eye is not particularly large. Even within the Collembola, most cave species are related to surface species that have even fewer eyes. The three major North American cave genera, *Pseudosinella, Sinella,* and *Tomocerus,* have epigean species with less than eight corneas (see Table 4–1). The situation is complicated because *Pseudosinella* and *Sinella* are somewhat artificial genera, defined as reduced-eyed *Lepidocyrtus* and *Entomobrya,* respectively. However, the closest surface ancestors of *Pseudosinella* and *Sinella* are other *Pseudosinella* and *Sinella,* not *Lepidocyrtus* and *Entomobrya*.

For many cave-limited species, the level of eye development of their surface ancestors is unknown. However, there is considerable variation in the amount of eye degeneration among cave-limited species, because of differences in surface ancestors, time of isolation in caves, or intensity of selection. This variation is exemplified by beetles in North American caves north of Mexico. Among the predaceous carabid

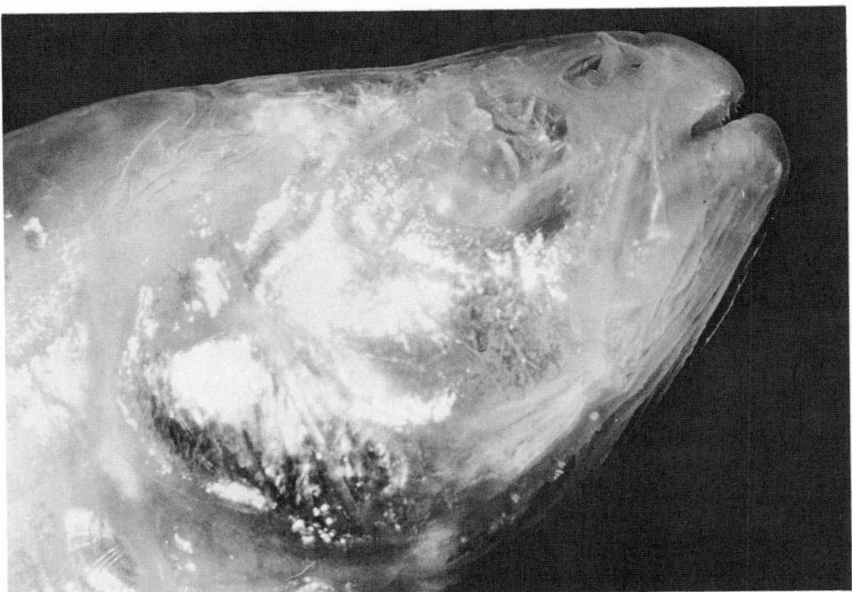

Figure 4–1 Surface-dwelling (*top*) and cave (*bottom*) characin fish, *Astyanax mexicanus*. (Photos courtesy of Dr. Robert W. Mitchell, Department of Biology, Texas Tech University, Lubbock, Texas.)

beetles, some genera, such as *Neaphaenops* and *Xenotrechus*, have no external manifestations of eyes (Barr and Krekeler 1967, Barr 1979), as is true for most species in the very large genus *Pseudanophthalmus*. However, a few Appalachian cave species have vestigial eyespots, and one, *P. petrunkevitchi*, retains a pigmented eye spot (Barr 1965). By contrast, all species in the *Rhadine subterranea* group, which inhabits Texas caves, retain a vestigial eye spot (Barr 1974a). The small, pale areolae range in size from 0.12 mm × 0.18 mm in *R. persephone* to only a tiny scar in fully sclerotized *R. noctivaga*. Among pselaphid beetles, all species of *Arianops*, both cave and soil inhabitants, lack eyes (Barr 1974b). The genus *Batrisodes* presents a very different picture. In the cave-limited species, the eyes of the female are a third to half the size of those of males (Park 1951). In the scavenger leiodid beetles in the genus *Ptomaphagus*, all cave-limited species have at least an eye vestige, with considerable interspecific variation (Peck 1973a).

In general, cave-limited amphipods from the same area show greater eye reduction than beetles. All species of *Stygobromus* and *Allocrangonyx* lack external signs of eyes (Holsinger 1971, 1978), including those *Stygobromus* found in deep lakes in the western United States (Holsinger 1974). Most cave-limited *Crangonyx* also lack eyes, but Holsinger (1972) reports that occasional individuals of *C. antennatus* have a few pigment flecks.

The Importance of Isolation

There is a general consensus among cave biologists that differentiation of cave populations, especially with regard to regressive evolution, cannot occur when there is gene exchange with surface populations. Probably the most common cause of isolation of a cave population is the extinction of a surface population through climatic or geological change, such as regional warming or uplift. Several lines of evidence point to the critical role of isolation. Troglophilic Collembola have approximately the same number of corneas per eye as epigean Collembola (Table 4–1); only troglobitic Collembola show consistent reduction in the number of corneas. In no troglophilic Collembola are there consistent differences between surface and cave populations in cornea number (Christiansen and Bellinger 1980–81). I know of no cases of a cline in eye degeneration due to genetic differences from a surface habitat into a cave. Finally, few cave populations with extensive eye degeneration can still interbreed with surface populations. One notable

Table 4–1 Number of corneas for Nearctic Collembola species, not including introduced species. (Data from Christiansen and Bellinger 1980–81.)

Genus	Ecological classification	No. species	Mean no. corneas	Range
Sinella	Epigean	1	2.0	—
	Troglophile	9	1.1	0–3
	Troglobite	5	0.0	—
Pseudosinella	Epigean	4	1.1	0–5
	Troglophile	6	2.2	0–6
	Troglobite	12	1.0	0–6
Tomocerus	Epigean	4	5.2	3–6
	Troglophile	8	6.0	—
	Troglobite	3	0.5	0–2

exception is *Astyanax mexicanus,* discussed in more detail below. Cave populations of this species were isolated from surface populations for considerable periods of time as a result of subterranean stream capture (Mitchell, Russell, and Elliott 1977).

In north temperate areas, most terrestrial cave species were isolated in caves during one of the warming periods of Pleistocene interglacials (Barr 1960, 1965). Which interglacial is more difficult to determine, and different groups may have been isolated during different interglacials (see chapter 7). If we take the beginning of the Holocene as a minimum estimate of time of isolation, then terrestrial populations have been isolated in caves for at least 10,000 years.

For aquatic cave species in north temperate areas, estimates of time of isolation vary with different taxonomic groups, but current opinion is that aquatic species have been isolated in caves longer than terrestrial species. Hobbs and Barr (1972) suggest that epigean ancestors of the *pellucidus* section of the crayfish genus *Orconectes* went extinct in the late Pliocene or early Pleistocene, about 2×10^6 years ago. Holsinger (1978) suggests an even earlier isolation for the amphipod genus *Stygobromus* in caves and other hypogean habitats and states that "their invasion of subterranean waters for the most part is probably unrelated to the climatic vicissitudes of the Pleistocene" (p. 141).

Two other groups of cave organisms are worth noting, because they seem to have been isolated relatively recently and yet show considerable regressive evolution. Cave populations of the Mexican characin *Astyanax mexicanus* were probably isolated in the late Pleistocene or even more recently—less than 10,000 years ago (Mitchell et al. 1977). The remarkably diverse Hawaiian cave fauna occurs in lava tubes that are between 90 and 20,000 years old, but Howarth (1972) believes the species are able to disperse between lava tubes, so that regressive evolution may have been occurring during much of the existence of the island itself, that is, a million years for Hawaii, the youngest island. Howarth (1980) also suggests that in this case cave populations are much larger than surface populations and that therefore isolation is unnecessary, because cave populations will swamp epigean populations rather than vice verse.

Within-Population Variation

Some populations that are apparently rather recently isolated in caves show great variation in eye development and pigmentation. The classic examples are populations of the isopod *Asellus aquaticus* in Yugoslavian caves. Kosswig (1948, 1965) found considerable variation in eye development and pigmentation (Fig. 4–2). He attributes this to relaxed selection, but introgressive hybridization with a surface population may be the cause of variability in the Grand Dôme population, which has a bimodal distribution of body pigmentation.

There is considerable variation in the amount of regressive evolution in different cave populations of *Astyanax mexicanus*. In most caves these differences are probably due to differences in the length of time of isolation (Wilkens 1976). Analysis of eyeball sizes of individuals from La Cueva de El Pachón and La Cueva de los Sabinos (Fig. 4–3) indicates considerable variation within populations (Wilkens 1971). The variation in the La Cueva Chica population is probably due to hybridization with the surface ancestor after isolation in caves. Present cave populations are still fully interfertile with surface populations (Şadoglu 1957). Because of this high level of hybridization, it is unfortunate that the characins of La Cueva Chica are the best-known cave characins. Mitchell, Russell, and Elliott (1977) suggest that high food levels in the cave allow surface forms to persist. Vandel (1964) gives other examples of variation in eye degeneration and pigment loss in cave species.

The crayfish *Procambarus simulans simulans* studied by Maguire (1961) in Longhorn Caverns, Texas, shows a cline in pigmentation. Crayfish deep in the cave are almost completely white, while those collected upstream and closer to the entrance are progressively more pigmented. But environmental factors can have a large effect on pigmen-

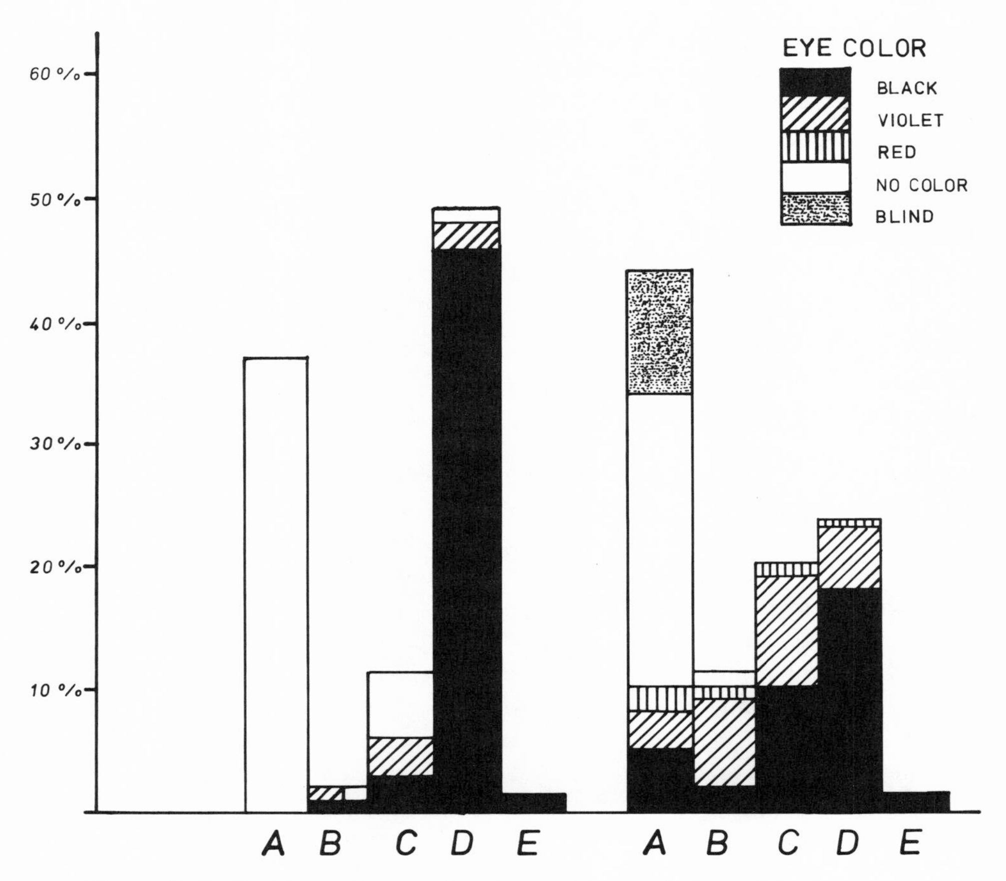

Figure 4–2 Variability in eye color observed in populations of *Asellus aquaticus* from Grand Dôme (*left*) and 1a Grotte Noire (*right*) in Yugoslavia. The percentages given are relative to the entire population for each location. Samples *A* through *E* are arranged by increasing body pigmentation, *A* being depigmented and *E* being heavily pigmented. (From Kosswig 1965.)

tation. Maguire showed that *P. simulans* from deep in the cave and from a nearby surface pond had the same color when grown in the same conditions. Crayfish reared with food containing algae were much darker than those reared with food lacking algae. Animals raised in the dark were also somewhat lighter than animals raised in the light. Whether the environmental factor is nutritional, especially in relation to carotenoid availability, as Maguire found, or light, which may be required for pigment formation (Beatty 1949), depends on the species. Gooch and Hetrick (1979) found no allozyme differences between spring and cave populations of the amphipod *Gammarus minus* in West

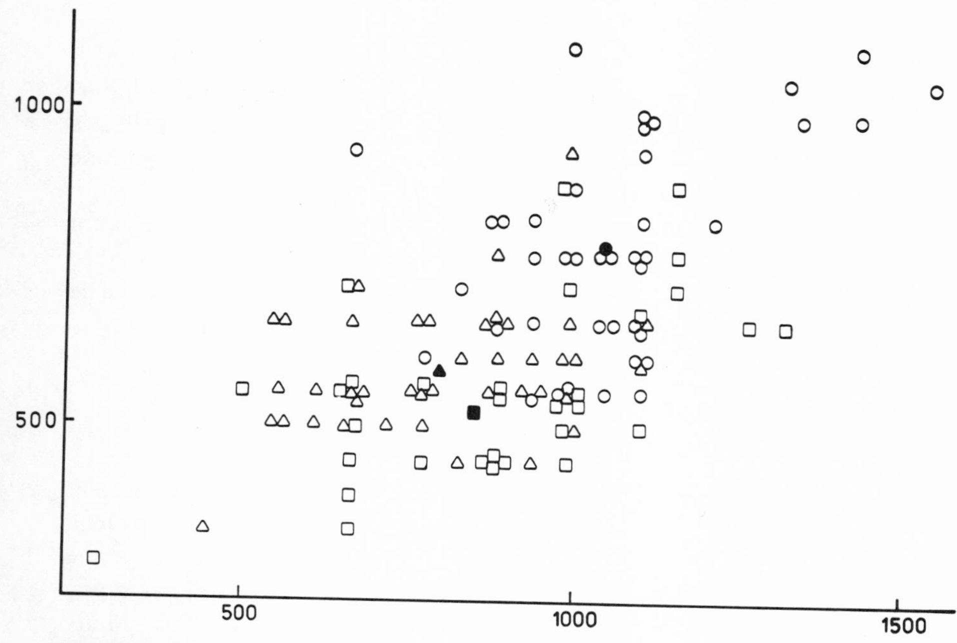

Figure 4–3 Eyeball sizes of *A. mexicanus*. Triangles represent those from Sabinos; squares, from Pachon; and circles are F_1 hybrids. Mean values are shown in black. The ordinate is the transverse diameter and the abscissa is the proximodistal diameter. (From Wilkens 1971.)

Virginia even though the number of distinct facets approached zero in some cave populations and 40 in some spring populations.

The Genetic System

Information on the number of loci involved in eye and pigment loss is not extensive, but it is clear that multiple loci are involved (Wilkens 1980). Kosswig and Kosswig (1940) report that five loci are involved in pigment loss in *Asellus aquaticus*. Wilkens (1970, 1971) reports two to four loci involved in melanophore reduction in cave populations of *Astyanax mexicanus*. Eye degeneration in *Astyanax* involves more loci; Wilkens (1971, 1976) suggests ten or more. The minimum number of loci involved in change in eye diameter can be calculated using the formula (Falconer 1960):

$$K = \frac{D^2}{8\sigma_A{}^2} \qquad (4\text{-}1)$$

where K is the minimum number of loci, D is the parental difference in eye diameter, and $\sigma_A{}^2$ is F_2 variance. This gives an estimate of five loci as a minimum, which is certainly an underestimate. The most important point to emerge from Wilkens' very careful studies is that a number of mutated loci, rather than one or two main loci, determine the amount of eye degeneration.

Wilkens (1971) also showed that in different populations, different loci are involved. The increase in eye size of F_1 hybrids compared to the parental populations from La Cueva de El Pachón and La Cueva de los Sabinos (Fig. 4–3) indicates that the genes responsible for eye reduction are at different loci in the two populations.

Population Size

As will be seen below, population size has a large effect on the rate of regressive evolution as a result of neutral mutation. Only a very small portion of many cave populations is observable; numerous cave species are known from only two or three specimens, such as many cave pseudoscorpions in the Appalachians (Muchmore 1976). Many *Pseudanophthalmus* beetles in Appalachian caves are only slightly more common, judged by the number observed at any one visit to a cave. At the other end of the spectrum are obviously large populations in which hundreds of individuals can be observed, for example, the beetle *Neaphaenops tellkampfi* in Mammoth Cave. A few mark-recapture studies have been done with cave populations (Table 4–2). While a few populations are of the order of 10^4, most are probably 10^3 or less, especially in view of the fact that almost all the marking studies reported in the table were done on large populations. Invertebrate species associated with bat guano, which show little regressive evolution, usually have populations between 10^4 and 10^5 (Mitchell 1970).

The Challenge of Cave Characins

The best-studied case of regressive evolution in cave animals is the Mexican cave characin, *Astyanax mexicanus,* originally studied by Breder and his associates in the 1940s (see Gresser and Breder 1940). This case is difficult to explain by either selectionist or neutralist hypotheses. These fish exhibit extensive regressive evolution that happened several times independently when different surface stream populations were isolated in caves by stream piracy (Mitchell, Russell and Elliott 1977). There has been considerable reduction of both eyes and

Table 4–2 Mark-recapture population estimates for various cave species.

Species	Cave	Population size	Source
Pisces			
Astyanax mexicanus	El Cueva de El Pachón, Mexico	8,502	Mitchell, Russell, and Elliott 1977
Astyanax mexicanus	El Sótano de Yerbaniz, Mexico	8,671	Mitchell, Russell, and Elliott 1977
Decapoda			
Aviticambarus jonesi	Shelta Cave, Alabama	345	Cooper 1975
Aviticambarus sheltae	Shelta Cave	100	Cooper 1975
Orconectes australis australis	Shelta Cave	1,000	Cooper 1975
Orconectes inermis inermis	Pless Cave, Indiana	9,090	Hobbs 1973
Coleoptera			
Ptomaphagus loedingi	Barclay Cave, Alabama	813	Peck 1975b
Pseudanophthalmus tenuis	Murray Spring Cave, Indiana	1,998	Keith 1975

pigment (see Fig. 4–1). In some caves, eye loss has proceeded to the point where there is no optic nerve and hardly anything is left of the eye capsule (Breder 1944). Accompanying the reduction in eyes and pigment are changes in the skull, shown in Figure 4–4, and in the pineal gland. In the cave populations there is an ontogenetic regression of the outer segment organization of the pineal (Herwig 1976).

The extensive changes in *A. mexicanus* occurred in a relatively short time (approximately 10,000 years), with large populations (Table 4–2), and in caves with abundant food in the form of bat guano. As we will see, these facts create difficulties for both theories.

Neutral Mutation Hypothesis

Three principal objections have been raised to the hypothesis that regressive evolution is the result of the accumulation of selectively neutral mutations. First, the process takes too long (Barr 1968); second, mutations should appear that increase as well as decrease a structure (Prout 1964, Şadoglu 1967); and third, there are independent reasons to believe that selection is involved (Poulson 1963, Howarth 1980).

The question of independent evidence implicating selection will be discussed in the next section, but it should be noted that the evidence is ambiguous. The other two objections are potentially fatal to any neutral mutation hypothesis and must be considered carefully. In what follows, I will show that it is at least plausible that accumulation of selectively neutral mutations can account for regressive evolution.

Mutations that have both increased and decreased the size of structures are well known. An example of increase is the "ultrabithorax" allele in *Drosophila* that converts halteres to wings (Prout 1964). An example of decrease is the "vestigial" wings in *Drosophila*. However, there is some reason to believe that the frequency of mutations that reduce a structure is much higher than the frequency of those that increase a structure. Prout (1964) expresses doubts, but Wright (1964 p. 66) holds that "new alleles would on the average tend to bring about reduction of the organ after its maintenance had ceased to be an object of natural selection." There is usually a set of mutations that results in a nonfunctional enzyme, and this set is likely to be larger than the set that in some way "improves" enzyme kinetics. What is needed to resolve this question in relation to cave species is some quantitative information about the average effect of mutations on eye size or pigment development.

The question of time can be dealt with more fully. Based on esti-

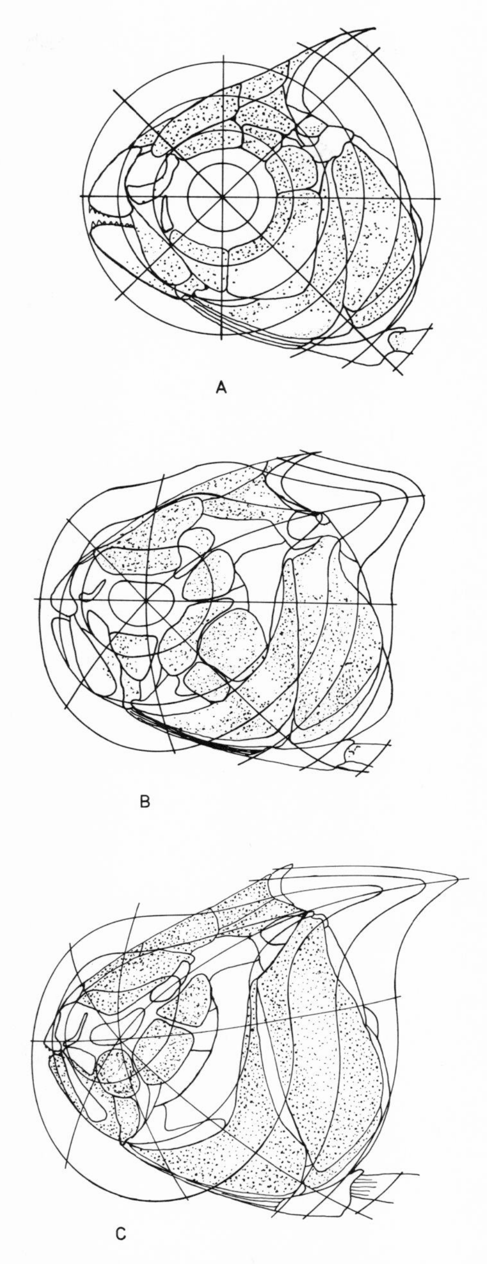

Figure 4–4 Transformation of (*A*) normal-eyed river *Astyanax mexicanus* into (*B*) La Cueva Chica *A. mexicanus* and (*C*) Cueva de los Sabinos *A. mexicanus,* using polar coordinates centered on the eye. This shows the skeletal consequences of eye loss. (From Breder 1944.)

mates of time of isolation in caves, there has been between 10^4 and 10^5 years for regressive evolution in terrestrial species and between 10^5 and 10^6 years in aquatic species. Given the longer generation time of aquatic species (see chapter 3), any process that takes more than 10^4 to 10^5 generations cannot explain regressive evolution. Most previous calculations of the time required for mutations to accumulate suffer from two defects: only a single locus was considered, and the effect of small population size was ignored.

The importance of including more than one locus is that, as Prout (1964) points out, it is no longer necessary for any particular allele to be fixed or even to come close to fixation. In the simplest case, consider a ten-locus system in which an allele at each locus completely blocks development. If each of these alleles is at a frequency of 0.70, the probability of an organism having the structure is only 6×10^{-6}. Since no allele has to become fixed, the time required is reduced. A more realistic model is that no single allele blocks development, but that some number of alleles, say n, at N loci are required to reduce the eye or pigment system to the observed level.

Table 4–3 shows some sample calculations of the time required for the accumulation of random mutations to produce degeneration. Time estimates were obtained as follows. First, the gene frequency, q_t, required to reduce the probability of an individual having fewer than n "reducing" alleles to $(10\ N_e)^{-1}$, where N_e is effective population size, was found by trial and error, using the first n terms of the binomial formula:

$$\sum_{j=0}^{n-1} \frac{N!}{j!(N-j)!} q^j(1-q)^{N-j} \tag{4-2}$$

The initial frequency, q_o, was set to $1/2N_e$. A reducing allele was assumed to appear the same time at each of the n loci with this assumption. The time required to change the frequency of the reducing allele from q_o to q_t, assuming irreversible mutation, is

$$t = \frac{\ln[(1-q_t)/(1-q_o)]}{\ln(1-\mu)} \tag{4-3}$$

where μ is mutation rate. From the values in Table 4–3 it is clear that if mutation rates are high ($\mu = 10^{-4}$ or higher), if the total number of loci is large ($N = 20$), or if the number of reducing alleles required is small ($n = 3$), then the accumulation of neutral mutations is rapid enough to

Table 4–3 Estimates of time in generations for a reducing allele to change in frequency from q_o to q_t. N_e is effective population size, N is the number of loci, n is the number of reducing alleles, μ is the mutation rate, and age is the average age, in generations, of an allele at frequency q_t. See text for details.

					Time			
N_e	N	n	q_o	q_t	$\mu = 10^{-4}$	$\mu = 10^{-5}$	$\mu = 10^{-6}$	Age
10^3	10	3	5×10^{-4}	0.8	2×10^4	2×10^5	2×10^6	4×10^3
10^4	10	3	5×10^{-5}	0.9	2×10^4	2×10^5	2×10^6	4×10^4
10^3	10	5	5×10^{-4}	0.9	2×10^4	2×10^5	2×10^6	4×10^3
10^4	10	5	5×10^{-5}	0.95	3×10^4	3×10^5	3×10^6	4×10^4
10^3	20	3	5×10^{-4}	0.55	8×10^3	8×10^4	8×10^5	3×10^3
10^4	20	3	5×10^{-5}	0.60	9×10^3	9×10^4	9×10^5	3×10^4
10^3	20	5	5×10^{-4}	0.65	9×10^3	9×10^4	9×10^5	3×10^3
10^4	20	5	5×10^{-5}	0.75	10^4	10^5	10^6	3×10^4

account for regressive evolution. It is a matter of pure speculation whether mutation rates and the number of loci involved in regressive evolution are within this "window."

In any event, these calculations ignore the effect of population size, N_e, except to set the initial gene frequency. A more realistic time estimate would take into account the effect of population size on the rate of change in gene frequency. The concept of the average age of an allele (Nei 1975) is useful in this regard. The average age of an allele, at frequency q, is the time in generations since it first appeared, given that it will eventually be fixed in the population. For a neutral allele this is

$$-[4N_e q/(1 - q)] \ln q \tag{4-4}$$

These time estimates are shown in the last column of Table 4–3. They are all much less than the time available for regressive evolution, but they do not include the time until the first appearance of an allele that will eventually be fixed. The probability that a neutral mutation will be fixed is $1/2N_e$ (Kimura and Ohta 1971). If the mutation rate is μ, then $1/2N_e$ of $2N_e\mu$ mutations will be fixed in each generation. If μ is 10^{-5}, the time between appearances of neutral mutations that will become fixed is 10^5 generations, which is as long as or longer than the time available for regressive evolution. However, since the model assumes that multiple loci are involved, the variance in time between appearances of neutral mutations that will become fixed is important and potentially will reduce the time considerably, especially if N (the number of loci) is large and n (the number of reducing alleles) is small. For weak selection, this waiting time can also be significant. The probability of fixing a slightly favored allele is also small, although not as small as the probability of fixing a neutral allele (Table 4–4).

Although the above calculations are incomplete and by necessity inconclusive until more information is available on mutation rates and number of loci, they do indicate that accumulation of neutral mutations is possibly the cause of regressive evolution. One can predict that if accumulation of neutral mutations is important, then those species that have been isolated in caves for a longer time should show more regressive evolution than species isolated for a shorter time. The greater reduction in the eyes of cave amphipods compared to the eyes of beetles in eastern North American caves agrees with this prediction. Wilkens (1973) also presents data indicating that Yucatan cave teleosts show increasing eye reduction with increasing time in caves. Cave teleosts of marine origin show greater eye reduction than those of freshwater origin, for whom the isolating event is more recent. This is perhaps the strongest field evidence for neutral mutation theory.

Can neutral mutation explain the regressive evolution of cave *Astyanax mexicanus?* Using the approach outlined above, the question can be answered only if we know the number of loci involved and the mutation rate (see Table 4–3). An alternative approach utilizes Lande's (1976) elegant work on the rate of phenotypic evolution. If some metrical character (for technical reasons the log of the actual measurement is usually employed) is reduced by some quantity z after t generations, then the reduction may have occurred by genetic drift alone with a probability of at least 5 percent if

$$N_e < N^* = \frac{(1.65)^2 h^2 t}{(z/\sigma)^2} \tag{4-5}$$

where N_e is the effective population size, h^2 is the heritability, and σ is the standard deviation of the metrical trait being considered. The question asked above can be rephrased as "Is there sufficient phenotypic variation to account for the observed changes?" Using Wilkens' (1971) data on eye diameters, the denominator of equation 4-5 is 10^2. Assuming 10^4 generations, N^* is 270 h^2. With a heritability of 0.5, the effective population size cannot be much larger than 10^2 for genetic drift to account for the observed amount of regressive evolution. The observed

Table 4–4 Probability of fixation of a beneficial allele that is recessive. Probability of fixation is approximately $\sqrt{2s/\pi N}$, where s is the selective advantage. See Table 4–3 for the meaning of other symbols. (From Nei 1975.)

s	$N_e = 10^3$	$N_e = 10^4$
0	5×10^{-4}	5×10^{-5}
0.001	8×10^{-4}	25×10^{-5}
0.01	25×10^{-4}	80×10^{-5}
0.1	80×10^{-4}	252×10^{-5}

population size is much larger (Table 4–2), but since N_e is the *harmonic* mean of population size, it will be close to the minimum population size. It is likely, although there is no supporting evidence, that population sizes of cave characins have gone through bottlenecks that make N_e much lower than the observed population size. Thus, as with neutral mutation theory, there is insufficient information to determine whether Lande's theory of genetic drift can account for the observed changes.

Selection

Pleiotropy Several authors, especially Wright (1964) and Barr (1968), have ascribed to pleiotropy an essential role in regressive evolution. The usual definition of pleiotropy is "the phenomenon of a single gene being responsible for a number of distinct and seemingly unrelated phenotypic effects" (King 1974), and by this definition pleiotropy can have several effects on regressive evolution. Perhaps the most obvious effect is the persistence of rudiments. If a mutation that causes a further reduction in a reduced structure, for example the optic lobe of amblyopsid fish (see Fig. 2–10), causes other deleterious effects, the vestigial structure persists because of pleiotropy. In this case the pleiotropic effects are positively correlated, in the sense that dysfunction in early eye development causes dysfunction in other physiological processes.

But both Wright and Barr implicate pleiotropy in the process of regressive evolution itself. Consider a mutation that affects the eye. If it reduces the eye and concomitantly improves function in other physiological processes, then the eye structure will be reduced by selection favoring the other phenotypic effects of the mutation. The critical question is whether a mutation that reduces the eye will improve other physiological functions. The bacterial genetics literature provides cases of negatively correlated pleiotropic functions. For example, Lin and his co-workers (Lin, Hacking, and Aguilar 1976; Wu, Lin, and Tanaka 1968) found that mutants of the mammalian gut bacterium *Klebsiella aerogenes* that were able to utilize xylitol did so at the expense of ribitol utilization. Besides structural changes at the ribitol dehydrogenase locus, there were alterations in repressor proteins.

Whether such changes are important in regressive evolution is unknown, and indeed the most frustrating aspect of the pleiotropy hypothesis is that it seems both important and untestable. However, as presented by Wright and Barr, the hypothesis clearly involves selection rather than mutation as the dominant factor. Furthermore, the experiments of Lin and his co-workers indicate that the hypothesis is feasible. Perhaps the best evidence in its favor is that with pleiotropy regressive evolution would proceed faster than it would by the accumulation of neutral mutations.

Energy Economy Aside from pleiotropy, energy economy is the most commonly invoked selective agent involved in regressive evolution of cave animals. The argument is simplicity itself. Since many cave organisms are strongly food limited (see chapters 2 and 3), a mutation that reduces a useless structure, such as the eye, will have a selective advantage because it saves energy. There is indirect evidence to support the hypothesis. According to this hypothesis, the fact that female *Batrisodes* beetles have smaller eyes than the males (Park 1951) can be explained because energy demands on females are greater than on males. The energy-economy hypothesis states that eyed, pigmented species isolated in caves have shown little regressive evolution because they are not strongly food limited. Mitchell's (1969) data from Texas caves indicates that eyed rhadinid beetles are usually found in caves with extensive guano and less limited food, while the eyeless beetles tend to be found in caves without guano (Table 4–5).

Table 4–5 Frequency of occurrence of eyed and eyeless rhadinid beetles in guano and nonguano caves in Texas. The number of caves is given in parentheses. (Data from Mitchell 1969.)

Cave type	Eyed species	Eyeless species
Guano	0.35 (16)	0.05 (2)
Nonguano	0.65 (30)	0.95 (41)

Why then is the energy-economy hypothesis not universally accepted? There are several reasons. If the hypothesis is correct, then species at higher trophic levels, where energy is likely to be a more limiting factor, should show more regressive evolution, but there is no evidence for this. Breder (1953) argues against energy economy because the African cave fish *Caecobarbus geertsii* and the Mexican cave fish *Astyanax mexicanus* have similar levels of eye degeneration, even though *Caecobarbus* is probably much more food limited than *Astyanax* and has been isolated in caves longer, judging by the absence of a putative ancestor within the geographic range of *Caecobarbus*. Mitchell's data on rhadinid beetles (Table 4–5) may also have an alternate explanation. Poulson (personal communication) points out that guano caves and nonguano caves tend to be in different geographic areas and suggests that eyed rhadinids may be more recently isolated rather than under a different selective regime.

A more fundamental objection concerns the real energetic importance of structures like pigment and eyes. No one has calculated the energetic cost of making an eye or pigment, nor even what proportion of biomass or calories of an organism is used by eyes or pigment. If the energy costs are very small, selective differentials will also be very small, so small that the dynamics of the different alleles may be very similar to those of neutral mutations (see Table 4–4). Whatever the selective values, one would expect those structures that tie up the most energy to disappear first if the additive genetic variance is the same. For example, eyes should disappear faster than pigment, but there is no evidence for this. Finally, Dykhuizen (1978), in a very elegant experiment involving the tryptophan loci in *E. coli,* showed that energy economy, which he was able to calculate directly, could not explain the gene frequency changes of a mutation that was unable to synthesize tryptophan. It is worth remembering that the neutral mutation hypothesis also has problems. In particular, smaller populations should have faster rates of regressive evolution, but there is no evidence for or against this.

Consider again the cave populations of *Astyanax mexicanus.* Assuming that no genetic drift occurs, Lande (1976) provides a way to determine the minimum selective differential required to account for the observed changes in a structure:

$$b = \pm\left[-2\ln\left(\sqrt{2\pi}\,\frac{|z|/\sigma}{h^2 t}\right)\right] \qquad (4\text{-}6)$$

where b is the number of standard deviations from the mean, and the other symbols are as in equation 4-5. This can be used to estimate selective differentials by referring to tables of a standard normal distribution. Using the same data as before on eye diameters of *Astyanax,* the minimum selective differential is 10^{-3}. While that is not large, it intuitively seems large for reduction in eye diameter in relatively food-rich caves.

Neither neutral mutation nor energy economy offers a completely satisfactory explanation for regressive evolution in *Astyanax*. Either the effective population size must be about 100, which seems too small, or the selective differential must be about 0.001, which seems too large. Pleiotropy can explain the patterns, but only because the pleiotropy hypothesis is very general, and it is not yet possible to make specific predictions about its consequences. Poulson (1981) suggests that both selection and mutation are involved in regressive evolution in general and that their relative importance depends on the structure involved (Table 4–6), selection being more important when the energy involved in a structure is great.

Summary and Retrospect

Our understanding of the causes of regressive evolution in cave animals is most unsatisfactory. Over the past several decades most cave biologists have assigned selection a major role in regressive evolution, in the form of either energy economy or of indirect effects of pleiotropy. Yet there are problems with both of these explanations. If energy economy is the general explanation, why don't the populations facing greater food scarcity, that is, populations at higher trophic levels, show greater regressive evolution? And why was energy economy not important in Dykhuizen's experiments with glucose-limited bacteria? If the indirect effect of pleiotropy is the general explanation, how can this hypothesis be made testable? How would it operate in a multilocus system where each allele has a small effect?

My own bias is that neutral mutations play a major role in regressive evolution, but I have no firm substantiation. This bias stems from the problems that exist with selection models and from the possibility that there has been sufficient time for neutral mutation processes to account for regressive evolution. While there are at least order-of-magnitude estimates for population sizes and for time since isolation, the same is not true for mutation rate, the number of loci, and the number of mutated loci. Until these are available, it is difficult to assess the plausibility of neutral mutation.

Table 4–6 Hypothesized role of selection and mutation in evolution of the cave spider *Anthrobia monmouthia*. (Considerably simplified from Poulson 1981.)

Mutation	Mutation and selection	Selection
Eye reduction	Exoskeleton thinning	Lower weight per unit length
Pigment reduction	Reduction in egg case (fewer layers of silk)	Larger egg size
Presence of epicuticular wax	Web reduction	

Part of the interest in regressive evolution is the possibility that it provides a way of measuring the evolutionary time since isolation in caves, in other words, that it is a crude sort of evolutionary clock. Both Poulson (1963) and Culver (1976) have used the amount of regressive evolution to judge the length of time various species have been isolated in caves. As a crude measure of time, such a procedure is probably justified if regressive evolution proceeds by neutral mutation or if the species are under similar selective regimes. In both cases the rate of evolution is constrained. If, on the other hand, the selective regimes are different, for example, in amount of food limitation, then regressive evolution is not even a crude measure of time.

The history of hypotheses about regressive evolution is replete with controversy and with hypotheses whose only adherents are their formulators. Barr (1968) gives a history of many of these ideas. One that has had many adherents is the Lamarckian and neo-Lamarckian hypothesis of disuse, which is worth considering because in many ways it is the most testable hypothesis yet generated. Darwin himself took a Lamarckian view of regressive evolution of cave animals: "As it is difficult to imagine that eyes, though useless, could be in any way injurious to animals living in darkness, I attribute their loss wholly to disuse" (Darwin 1859), a view that remained essentially unchanged in all subsequent editions of *The Origin of Species* (Peckham 1959).

Unlike the neo-Lamarckians, however, Darwin believed that loss through disuse would occur slowly: "By the time that an animal had reached, after numberless generations, the deepest recesses [of caves], disuse will on this view have more or less perfectly obliterated its eyes" (Darwin 1859). Neo-Lamarckians held that the loss of eyes through disuse would be very rapid: "Certain Aselli borne into caves . . . losing the stimulus of the light, begin to lose their eyes and the power of sight. The first step is the decrease in the number of facets and corresponding lenses and retinae; after a few generations—perhaps in four or five—the facets become reduced to only four or five; the eye is then useless; then all at once, perhaps after only two or three generations, there is a failure in forming images on the retina, and those complicated, elaborate structures, the optic ganglia and the optic nerves, suddenly break down and are absorbed, though the external eye still exists in a rudimentary state'' (Packard 1888).

Packard suggested two tests of neo-Lamarckian theory. First, according to the hypothesis, animals that spend their entire lives deep in caves should have nonfunctional, rudimentary eyes, and he went to considerable lengths to explain the presence of eyed species in caves. Most of such species that were known in Packard's time were found near cave entrances, and it is probably fair to say that none of the counterexamples were sufficiently well known at the time to falsify the theory. The second test that Packard proposed was to reverse regressive evolution by rearing cave species with rudimentary eyes in the light. Because of the rapidity of eye loss, reversal should have been possible within a few generations. The neo-Lamarckian hypothesis was falsifiable and obviously wrong. The current theories, especially selection theories, need to be reframed in a more testable way. A more critical need is for more data on the number of loci involved in a structure, phenotypic variability and change, heritability, population sizes, and time since isolation. Then the neutral mutation hypothesis can be tested more rigorously.

5 Allozyme Variation

The use of gel electrophoresis to estimate genetic variation has revolutionized population genetics but has not yet resolved the basic question of the field—how much genetic variation in fitness is there? Indeed, the simpler question of how much genetic variation there is has not been answered with complete satisfaction for any population. After the initial discovery of high levels of heterozygosity of soluble enzymes in *Drosophila* (Lewontin and Hubby 1966) and humans (Harris 1966), the main debate centered on whether this variation was maintained by selection or by the accumulation of neutral mutations. In spite of repeated claims to the contrary, both hypotheses are still viable, although not in their original form. This is not to say that in the last fifteen years gel electrophoresis has not increased our understanding of genetic variation and its causes. Rather it has shown that the causes of variation are much more difficult to decipher than originally suspected. More than in any other area of population biology, advances in this area depend on increased technical sophistication in extracting all of the genetic variation at a locus (see, for example, Singh, Lewontin, and Felton 1976). This poses a special problem for the discussion of genetic variation in cave organisms, which are difficult or impossible to breed in the lab. Also, researchers on the technological forefront do not usually work with cave organisms.

The Limitations of Gel Electrophoresis

To avoid both false optimism and complete despair concerning data on genetic variation in cave organisms, we must review the strengths and weaknesses of electrophoretic data in general. As Lewontin (1974) points out, any technique that enumerates genotypes requires that:

1. Phenotypic differences resulting from a single allelic substitution must be detectable as a difference between individuals.
2. Allelic substitutions at different loci must be distinguishable.
3. All, or a very large fraction of, allelic substitutions at a locus must be distinguishable.
4. The loci studied must be a random sample of genes with respect to variation and the amount of genetic variation.

It is a familiar refrain that techniques for analyzing genetic variation, such as cataloging visible mutations, do not satisfy these criteria, and that gel electrophoresis apparently does. For cave organisms, gel electrophoresis has the additional advantage of being the only genetic information available, except for *Astyanax*. There are no great catalogs of mutations, let alone stocks of mutants, available for cave organisms.

Much of the controversy surrounding gel electrophoresis can be summarized as follows. First, standard electrophoretic techniques extract only about a third of the variation at a locus (Singh, Lewontin and Felton 1976), thus violating Lewontin's third requirement. Second, soluble enzymes in general may be more variable than other proteins such as structural proteins (Leigh-Brown and Langley 1979). Thus estimates of genetic variation based on soluble enzymes may be too high. Third, because of the high within population variance in genetic variation, that is, some loci monomorphic and some highly polymorphic, statistical demonstration of differences among populations is very difficult (Lewontin 1974).

The seriousness of these problems depends on how the electrophoretic data are used. Consider first the problem of variation within populations. Two measures are commonly used. The first is average heterozygosity, H, the average proportion of loci heterozygous per individual:

$$\frac{\sum_{j=1}^{n}\left(1-\sum_{i} p_{ij}^{2}\right)}{n} = H \qquad (5\text{-}1)$$

where p_{ij} is the frequency of allele i at locus j. The second measure is the proportion of loci polymorphic, P, the fraction of loci that are polymorphic in a population. Using more sensitive electrophoretic techniques, Singh (1979) has shown that loci originally thought to be monomorphic remain monomorphic, and that some, but not all, polymorphic loci are much more polymorphic than originally thought. Thus P remains unchanged and H increases, but not by a large amount.[1]

The real importance of the discovery of additional variation is the challenge it presents to selectionist arguments, its impact on statistical tests of neutralist models (Ewens and Feldman 1976), and the large effect it has on measures of genetic distance (Coyne 1976). Using two-dimensional gel electrophoresis, Leigh, Brown, and Langley (1979) have claimed that structural proteins as a group are much less variable than soluble proteins, reducing genome-wide H and P by perhaps an order of magnitude. While this claim is open to question, its consequences relate to the overall likelihood that neutral mutation or selection explains observed heterozygosity in general. It does not affect comparisons among species.

The third question, the statistical demonstration of differences among populations, is the most serious. Large standard errors of both H and P are intrinsic because most loci are monomorphic and some are highly polymorphic (Lewontin 1974). Lewontin's discussion of this problem has had a curious but perhaps not unexpected effect: there has been little statistical analysis. On the surface, the problem is empirical—simply get estimates of variation at a large number of loci. But hidden in this are questions of the equivalence of loci and the equivalence of investigators. In order to use standard statistical tests such as analysis of variance, one must assume that different loci are independent and that each locus has an equal likelihood of variation. Independence of loci assumes the absence of linkage disequilibrium, and in the absence of chromosomal inversions or translocations, this may be justified (Charlesworth and Charlesworth 1973). The equal likelihood of variation is a much more serious problem. Selander (1976) documents that different enzymes have greatly differing amounts of variation, although the reasons are not always clear. When comparing species, one is usually comparing different enzymes as well, so that significant statistical differences may be due either to ecological and evolutionary differences or to intrinsic enzyme differences. The problem is compounded by the fact that different researchers tend to

1. Let L be the number of alleles, $\partial^2 H/\partial L^2 < 0$.

analyze different enzyme systems. In spite of these problems, and in a sense because of them, extensive use of t-tests and analysis of variance will be used.

Cave versus Epigean Populations

Several hypotheses have been advanced to account for possible differences in heterozygosity between cave and epigean populations. In general, neutralist hypotheses predict lower heterozygosity in cave populations because H is positively correlated with population size:

$$H = \frac{4N_e\mu}{4N_e\mu + 1} \tag{5-2}$$

where N_e is the effective population size and μ is the mutation rate (Kimura and Ohta 1971). A more realistic version would include the time that has elapsed since a population bottleneck, which is also positively correlated with heterozygosity (Soulé 1976), and the probability of fixation of slightly deleterious mutants (Maruyama and Kimura 1978). A parallel selectionist argument is that since caves tend to be less variable, both spatially and temporally, cave populations will show less heterozygosity (Nevo 1976). However, another selectionist argument predicts the opposite: since cave organisms have fewer competitors and predators, ecological release may allow habitat expansion, increased niche breadth, and increased heterozygosity (Dickson et al. 1979).

Estimates for heterozygosity and polymorphism for a variety of cave species are summarized in Table 5–1. The species are grouped by a combination of ecological and taxonomic criteria. Among the wide variety of cave species included are cave crickets that periodically leave caves to feed (trogloxenes), predatory cave-limited beetles, marine relict isopods, and cave-limited fish.

The most basic question to be asked is whether cave species show any consistent differences from surface species. An attempt at appropriate comparisons is summarized in Table 5–2, in which terrestrial cave arthropods are compared to *Drosophila*, aquatic cave arthropods to marine invertebrates, and cave fish to epigean fish. Terrestrial cave arthropods, both trogloxenes and troglobites, have lower heterozygosity per individual and lower frequency of polymorphism than *Drosophila*, but the only statistically significant difference is the lower polymorphism of terrestrial troglobites. Both troglophilic and troglobitic

Table 5–1 For cave populations, estimates of the mean of average heterozygosity ($\bar{H}$) predicted by Hardy-Weinberg proportions, proportion of loci polymorphic ($\bar{P}$), and Nei's (1972) index of genetic similarity ($\bar{I}$). Reports of data on less than eight loci and of combined populations are not included.

Species	No. populations	No. loci	$\bar{H}$	$\bar{P}$	$\bar{I}$	Source
Terrestrial trogloxenic arthropods						
Troglophilus cavicola	2	16	.082	.469	.985	Sbordoni et al. 1981
Troglophilus andreinii	3	14–16	.163	.670	.928	Sbordoni et al. 1981
Ceuthophilus gracilipes	8	26	.028	.085	—	Cockley, Gooch, and Weston 1977
Terrestrial troglobitic arthropods						
Scoterpes copei	2	10	.032	.150	.217	Laing, Carmody, and Peck 1976a
Ptomaphagus hirtus	6	13	.056	.154	.810	Laing, Carmody, and Peck 1976b
Neaphaenops tellkampfi tellkampfi	8	13	.192	.470	.980	Turanchik and Kane 1979
Speonomus delarouzeei	7	12	.101	.489	.756	Delay et al. 1980
Aquatic troglophilic arthropods						
Gammarus minus	11	13	.025	.091	.959	Gooch and Hetrick 1979
Aquatic troglobitic arthropods						
Monolistra boldorii	1	18	.303	.714	—	Sbordoni et al. 1980
Monolistra berica	1	18	.321	.714	—	Sbordoni et al. 1980
Monolistra caeca	1	18	.277	.786	—	Sbordoni et al. 1980
Crangonyx antennatus	6	8	.118	.250	.920	Dickson et al. 1979
Troglophilic fish						
Astyanax mexicanus	3	17	.036	.137	.834[a]	Avise and Selander 1972
Chologaster agassizi	10	19–22	.028	—	.895	Swofford, Branson, and Sievert 1980
Troglobitic fish						
Typhlichthys subterraneus	13	19–22	.019	—	.679	Swofford, Branson, and Sievert 1980
Amblyopsis spelaea	3	19–22	.000	.000	1.000	Swofford, Branson, and Sievert 1980
Amblyopsis rosae	2	19–22	.006	—	.854	Swofford, Branson, and Sievert 1980

a. Rogers' Index.

Table 5–2 Comparison of average heterozygosity ($\bar{H}$) and polymorphism ($\bar{P}$) for cave and surface species. Heterozygosity and polymorphism for each cave species was computed by finding the average for each population (see Table 5–2). Standard errors are shown following the means. N is the number of species. (Data for surface species are from Selander 1976.)

Cave species				Surface species				T-test			
								$\bar{H}$		$\bar{P}$	
Group	N	$\bar{H}$	$\bar{P}$	Group	N	$\bar{H}$	$\bar{P}$	t	Prob.	t	Prob.
Terrestrial trogloxenic arthropods	3	.091 ± .039	.41 ± .17	*Drosophila*	28	.151 ± .010	.53 ± .03	1.826	N.S.	1.142	N.S.
Terrestrial troglobitic arthropods	4	.095 ± .035	.32 ± .04	*Drosophila*	28	.151 ± .010	.53 ± .03	1.911	N.S.	2.439	< 0.05
Aquatic troglobitic arthropods	4	.255 ± .046	.62 ± .12	Marine invertebrates	9	.147 ± .019	.59 ± .08	2.637	< 0.05	0.218	N.S.
Aquatic troglophilic fish	2	.032 ± .004	—	Fish	14	.078 ± .012	—	1.411	N.S.	—	—
Aquatic troglobitic fish	3	.008 ± .006	—	Fish	14	.078 ± .012	—	2.643	< 0.05	—	—

fish follow a similar pattern, with heterozygosity lower in cave species than in epigean species. The only significant difference is in heterozygosity between troglobitic and epigean fish. The pattern of lower heterozygosity and polymorphism in cave organisms is reversed in aquatic invertebrates. Aquatic cave invertebrates show both higher polymorphism and higher heterozygosity than marine invertebrates, and the difference in heterozygosity is significant.

It is clear from Table 5–2 that cave species do not show consistent differences in the same direction in heterozygosity and polymorphism compared to surface species. The observed differences may or may not hold up when more extensive data are available, but since the differences are not in the same direction, it is unlikely that the pattern will become any simpler.

Differences among Cave Species

It is also clear from the data in Tables 5–1 and 5–2 that there are considerable differences among cave species. One particularly illuminating example is from White's Cave and Great Onyx Cave in Kentucky, which are most likely a single cave hydrologically (Laing, Carmody, and Peck 1976b). Estimates of genetic variation are available for three species of troglobites in the cave: the milliped *Scoterpes copei* (Laing et al. 1976a), the leiodid beetle *Ptomaphagus hirtus* (Laing et al. 1976b), and the carabid beetle *Neaphaenops tellkampfi* (Turanchik and Kane 1979). Both the proportion of loci polymorphic and the average heterozygosity vary greatly, with *Neaphaenops* having the highest H and P (Table 5–3). Assuming independent loci and equivalence of loci, an analysis of variance can be employed (Table 5–4). The *a posteriori* HSD test indicates that *Neaphaenops* differs from the other two species, for reasons that will be considered later. The important point is that there are differences among cave species.

Table 5–3 Genetic variation in *Neaphaenops tellkampfi, Ptomaphagus hirtus,* and *Scoterpes copei* from White's/Great Onyx Cave, Kentucky. (Data from Laing, Carmody, and Peck 1976a,b; Turanchik and Kane 1979.)

Species	No. individuals	No. loci	$\bar{H} \pm$ S.E.	$\bar{P} \pm$ S.E.
S. copei	22	10	.008 ± .008	.10 ± .10
P. hirtus	50–78	13	.133 ± .042	.15 ± .10
N. tellkampfi	21–46	13	.228 ± .063	.46 ± .14

Table 5–4 Analysis of variance of heterozygote frequencies (transformed by $\sin^{-1}\sqrt{p}$) of *N. tellkampfi, P. hirtus,* and *S. copei* from White's/Great Onyx Cave in Kentucky.

	df	SS	MS	F	p
Factor	2	1815.0	907.5	3.94	< 0.025
Error	33	7592.5	230.1		
Total	35	9407.5			

Explanations

Although no generalizations are possible or even justified, particular studies provide strong evidence for both neutralist and selectionist arguments. Two versions of the neutralist hypothesis have received support:

1. Small effective population sizes coupled with genetic drift result in low genetic variation in cave populations.
2. Differences in time since a genetic bottleneck account for differences in genetic variation among cave populations.

Likewise, two versions of the selectionist hypothesis have received support:

1. The relative environmental uniformity of caves results in low heterozygosity.
2. Reduced species diversity allows ecological release and high heterozygosity.

For particular species, a case can be made for each hypothesis.

Small Populations and Genetic Drift The most frequently invoked explanation of genetic variation in cave organisms is that small populations and little gene exchange among populations result in low heterozygosity and polymorphism and in fixation at different alleles in different populations. At least in theory, this hypothesis can be distinguished from the selectionist argument of relative environmental uniformity, which predicts that the same alleles should be fixed in different populations.

At this point a digression about the measurement of genetic similarity among populations is necessary. The most commonly used measure is Nei's index I, defined as follows (Nei 1972). If x_i and y_i are the frequencies of the ith alleles in populations x and y, the genetic identity between the two populations is, at locus j:

$$I_j = \frac{\Sigma x_i y_i}{(\Sigma x_i^2 \Sigma y_i^2)^{1/2}} \tag{5-3}$$

and overall identity is

$$I = \frac{J_{xy}}{(J_x J_y)^{1/2}} \tag{5-4}$$

where J_x, J_y, and J_{xy} are the means over all loci of Σx_i^2, Σy_i^2, and $\Sigma x_i y_i$, respectively. The small population, genetic drift argument predicts a low I, while the selection in a uniform environment argument predicts a high I. The practical problem is that, unlike H and P, I is very sensitive to the presence of undetected variation (Coyne 1976), and apparently high genetic identities may therefore be spurious.

Genetic drift in small populations has been invoked for many of the species listed in Table 5–1, including *Ceuthophilus gracilipes, Scoterpes copei, Ptomaphagus hirtus, Gammarus minus,* and *Astyanax mexicanus,* whose mean heterozygosities range from 0.025 to 0.056. Using Maruyama and Kimura's (1978) formula for the expected heterozygosities with neutral and slightly deleterious mutants, one can determine the mutation rates needed to produce the observed heterozygosity:

$$\mu = \frac{[(1 - H)^{-2} - 1]}{8N_e} \tag{5-5}$$

where H is heterozygosity, N_e the effective population size, and μ the mutation rate. For population sizes between 10^2 and 10^4, mutation rates required to explain observed heterozygosities range from 3×10^{-4} to 5×10^{-7} (Table 5–5). For population sizes of 10^3 or larger, the required mutation rates are at least not unrealistic.

Two studies are especially difficult to explain by any other hypothesis. In the first, two populations of the milliped *Scoterpes copei* in Kentucky caves had no alleles in common at eight of ten loci studied (Laing et al. 1976a), and heterozygosity was very low in both populations (Table 5–6). In the second study, that of Avise and Selander (1972) on cave and surface populations of *Astyanax mexicanus,* the findings are in agreement with several predictions of the genetic drift

Table 5–5 Mutation rates required to obtain a given heterozygosity (H) for a range of effective population sizes, using Maruyama and Kimura's (1978) formula (see text).

	Effective population size (N_e)		
Heterozygosity (H)	10^2	10^3	10^4
0.02	5×10^{-5}	5×10^{-6}	5×10^{-7}
0.05	1×10^{-4}	1×10^{-5}	1×10^{-6}
0.10	3×10^{-4}	3×10^{-5}	3×10^{-6}

model. First, the cave populations had lower heterozygosity ($\bar{H}$ = 0.036) than the surface populations ($\bar{H}$ = 0.110). A nested analysis of variance of their data, using expected heterozygote frequencies (transformed by $\sin^{-1}\sqrt{H_i}$, where H_i is the expected frequency of heterozygotes under Hardy-Weinberg mating at locus i) indicates that surface populations do not differ among themselves, and neither do cave populations. However, cave populations are significantly different from surface populations (F = 13.8, df = 1.7, $P < 0.01$). Second, mutation rates required to produce the heterozygosities observed are not unreasonably large (see Table 5–5). Third, Rogers' coefficient of genetic similarity, which is similar to Nei's index, is greater in surface populations (.958 ± .003) than in cave populations (.835 ± .040), indicating greater differentiation of cave populations.

Time since Genetic Bottleneck It is very clear that not all the data in Table 5–1 can be explained by genetic drift in small populations, especially populations with high heterozygosities. The mutation rates needed to account for high heterozygosities in small populations are unreasonably large (Sbordoni et al. 1981). The idea that an important determinant of heterozygosity (see Soulé 1976) is length of time since a genetic bottleneck has been championed by Sbordoni and his associates (1980, 1981) and Delay (1980). Their most convincing evidence for the importance of time since colonization of caves is that the heterozygosity of the cricket *Troglophilus cavicola* population from Marmo Cave in Italy, which is more recent in origin than other *Troglophilus* studied, had a lower heterozygosity (0.059) than conspecifics ($\bar{H}$ = 0.104) or *T. andreinii* ($\bar{H}$ = 0.163). However, if the effective population size of *Troglophilus* is under 1,000, as Sbordoni et al. (1981) claim, then the accumulation of neutral and slightly deleterious muta-

Table 5–6 Gene frequencies at ten loci in *Scoterpes copei* from two Kentucky caves. (From Laing et al. 1976a; reprinted by permission of the University of Chicago Press, © 1976, the University of Chicago.)

Locus	Allele	Gene frequency	
		White's Cave	State Trooper Cave
Esterase 1	a	—	1.00
	b	1.00	—
Esterase 3	a	—	0.95
	b	—	0.05
	c	0.96	—
	d	0.04	—
Esterase 4	a	—	1.00
	b	1.00	—
Esterase 5	a	—	0.62
	b	—	0.38
	c	1.00	—
Octanol dehydrogenase 1	a	1.00	—
	b	—	1.00
Octanol dehydrogenase 2	a	1.00	—
	b	—	1.00
Indophenol oxidase 1	a	1.00	—
	b	—	1.00
Indophenol oxidase 2	a	—	1.00
	b	1.00	—
Lactate dehydrogenase 2	a	1.00	1.00
Malate dehydrogenase 4	a	1.00	1.00

tions over time cannot explain their results. The calculations in Table 5–5 are for a steady state, and recent bottlenecks would further reduce diversity. Either their estimates of N_e are too low, or selection is involved. Effective population size for *Troglophilus* may be much higher, since they are not limited to caves and considerable gene exchange may occur. Since H increases with effective population size, this could explain their results. Alternatively, some form of balancing selection may be involved.

Selection in a Uniform Environment For those cave-limited species that experience food scarcity and have been subject to strong directional selection for metabolic economy, increased longevity, and low fecundity (see chapters 2 and 3), one might expect low genetic variation, resulting from both the reduced environmental variation of caves and from directional selection itself (Poulson and White 1969). Since different caves in an area have very similar environments and selective regimes, the genetic similarity between populations should be high. Surprisingly, only one study has supported this hypothesis—the preliminary study of Swofford, Branson, and Sievert (1980) on amblyopsid fishes. They found that heterozygosity declined with increasing cave adaptation, culminating in a mean heterozygosity of 0.006 for *Amblyopsis rosae* and 0.000 for *A. spelaea* (Table 5–1). This decline is unlikely to be due to reduced population size, since the populations of *Amblyopsis* are generally larger than those of *Typhlichthys* (Poulson 1969), but *Typhlichthys* has higher heterozygosity.

Ecological Release Another potential candidate for the previous hypothesis is the beetle *Neaphaenops tellkampfi,* which has either, depending on one's perspective, specialized on the eggs of the cave cricket, *Hadenoecus subterraneus* (see chapter 6) or expanded its niche to include them. In any case, it is a highly modified cave beetle that is a major predator in the terrestrial community in many Kentucky caves. Extensive studies by Kane and his colleagues (Giuseffi, Kane, and Duggleby 1978; Turanchik and Kane 1979; Kane 1981; Brunner and Kane 1981) have shown that rather than low heterozygosity, *Neaphaenops* has one of the highest heterozygosities recorded for cave animals (see Table 5–1). Kane attributes this to ecological release; *Neaphaenops* is able to expand its niche to include cricket eggs as well as small arthropods. When cricket eggs are available, roughly half the year, they are the bulk of their food (Kane and Poulson 1976). Evidence supporting the ecological release hypothesis is provided by differences in mean heterozygosity among the subspecies of *N. tellkampfi*. Only in the range of the subspecies *N. tellkampfi tellkampfi* is there usually the sandy habitat suitable for predation on eggs (Barr 1979). By the ecological release hypothesis, this subspecies should have the highest mean heterozygosity, which it does (Table 5–7).

Kane (1981) offers the following hypothesis to explain the differences in heterozygosity among *Scoterpes copei, Ptomaphagus hirtus,* and *Neaphaenops tellkampfi* (Table 5–4). Since all are cave limited, one might expect them to have similar levels of heterozygosity. *Ptomaphagus hirtus* is more opportunistic, with local populations developing on isolated patches of carrion and vertebrate feces. Both its small population size and the resource restriction may be important in explaining its low heterozygosity. The low heterozygosity of *S. copei* is

Table 5–7 Genetic variation in four subspecies of *Neaphaenops tellkampfi*, using the same seven loci for comparison. (From Brunner and Kane 1981.)

Subspecies	No. populations	$\bar{H}$
henroti	2	0.091
viator	3	0.082
meridionalis	2	0.137
tellkampfi	8	0.154

more difficult to explain. Unlike *P. hirtus*, it is not opportunistic but is apparently restricted to low-energy foods such as dispersed guano of *Hadenoecus subterraneus* and litter left by receding floodwaters. Kane also notes that it is less vagile than *N. tellkampfi*, which results in reduction of the effective population size.

The Ecological Importance of Heterozygosity

For the last two decades population biologists have hoped that population ecology and population genetics would become a unified whole. A glance at Table 5–2 shows how far away this goal remains. Whether troglophiles are really facultative cave dwellers, "troglobites in training," or a distinct adaptive strategy, they should differ consistently from troglobites. But they don't. There are at least three possible reasons for this discordance. First, the estimates of heterozygosity and polymorphism are not directly comparable among species, explained in part by the fact that different investigators report data for different enzyme systems.

The second reason for the discordance may be a "neoneutralist" explanation, in which effective population size plays a critical role (Maruyama and Kimura 1978). There is some indication that effective population size may be the major determinant of heterozygosity. For example, *N. tellkampfi* most likely has a larger effective population size than *S. copei* and *P. hirtus* (Table 5–4), and *N. tellkampfi tellkampfi* most likely has a larger effective population size than the other subspecies (Table 5–5).

Finally, the genetic data may be more reliable than the ecological data, and the genetic differences summarized in Table 5–1 may represent the results of selection rather than neutral mutation, consequently demonstrating the imperfection of our ecological understanding.

Genetic Differentiation in Terrestrial Populations

Somewhat surprisingly, a clearer pattern emerges when one considers genetic differences among populations, especially since Nei's index is very susceptible to the sensitivity of electrophoretic technique (Coyne 1976). Although Table 5–1 shows a wide range of Nei's genetic similarity values, some of this discrepancy is more apparent than real. Some similarity values are for populations from nearby caves; others are for populations from distant caves. Extensive studies by Kane and his colleagues (Turanchik and Kane 1979; Kane 1981; Brunner and Kane 1981) on *N. tellkampfi* and Peck and his colleagues (Laing, Carmody, and Peck 1976b) on *P. hirtus* from the Pennyroyal Plateau and adjacent upland to the south and north in Kentucky allows detailed comparison of two species with very different heterozygosities (Table 5–4).

In a recent study of *N. tellkampfi,* Barr (1979) distinguishes four subspecies, based on minute but consistent morphological criteria. Their range (Fig. 5–1) roughly corresponds to the range of *P. hirtus* (Laing, Carmody, and Peck 1976b). Except for *N. t. henroti,* which is isolated by a heavily faulted ridge with considerable thickness of sandstone, there are no obvious geographic barriers between subspecies (Barr 1979). For the moment, the reality of the subspecies will be accepted.

Within the range of the nominate subspecies *N. t. tellkampfi* the average Nei index is 0.980 ± 0.003 (Turanchik and Kane 1979). In this same area the average Nei index for *P. hirtus* is 0.874 ± 0.045 (Laing, Carmody, and Peck 1979). By way of comparison, different geographic populations of species in the *Drosophila willistoni* group have a Nei index of 0.970 ± 0.006; subspecies have a Nei index of 0.795 ± 0.013 (Avise 1976). It is not surprising that the Nei index of *P. hirtus* is lower than equivalent populations of *D. willistoni.* What is surprising is that the Nei index is not lower for both *P. hirtus* and *N. t. tellkampfi.* The size of the area under consideration is over 50 km by 30 km (Fig. 5–1), so restricting the analysis to the range of *N. t. tellkampfi* has not defined the problem away. Furthermore, the Green River cuts deeply through the cavernous limestones and forms an obvious dispersal barrier. But it apparently does not act as a dispersal barrier, for *P. hirtus* populations in Running Branch Cave on the north side of the river and in White's/Great Onyx Caves on the south side have a genetic identity of 0.954. Likewise, *N. t. tellkampfi* populations from the same caves have a genetic identity of 0.966. Either there is significant undetected variation (see Coyne 1976), or possibly there is dispersal, perhaps in air bubbles (Barr and Peck 1965).

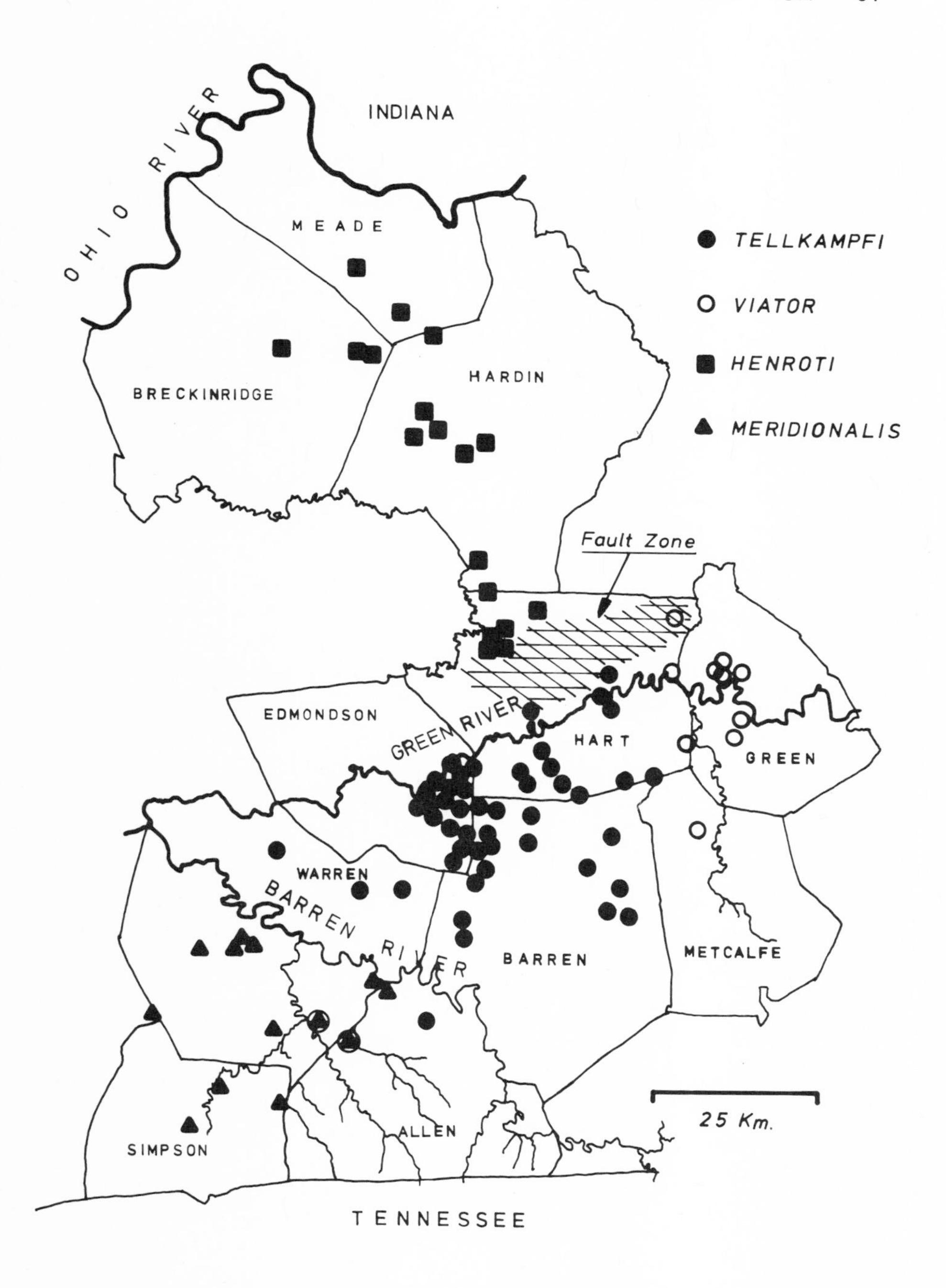

Figure 5–1 Distribution of the four subspecies of *Neaphaenops tellkampfi* in west central Kentucky. (From Barr 1979, courtesy of the American Museum of Natural History.)

Table 5–8 Genetic similarity values (I) for the four subspecies of *Neaphaenops tellkampfi*. Values on the principal diagonal represent within-subspecies similarities. Data are based on seven loci. (From Brunner and Kane 1981.)

	henroti	*viator*	*meridionalis*	*tellkampfi*
henroti	0.982	0.733	0.777	0.970
viator		0.940	0.758	0.674
meridionalis			0.958	0.758
tellkampfi				0.973
$\bar{I}$	0.823	0.723	0.764	0.801

Turning to relations among subspecies of *N. tellkampfi* and populations of *P. hirtus* in the range of the different *N. tellkampfi* subspecies, the average between-area Nei index is 0.797 ± 0.035 for *N. tellkampfi* (Brunner and Kane 1981) and 0.794 ± 0.032 for *P. hirtus* (Laing, Carmody, and Peck 1976b), very close to the average for subspecies in the *Drosophila willistoni* group (Avise 1976). The range of *P. hirtus* has been divided according to the ranges of the subspecies of *N. tellkampfi*, but *P. hirtus* nevertheless has genetic similarity values similar to geographic isolates in the process of speciation; therefore, many morphological species may contain several subspecies or even species, as is likely the case with *Scoterpes* (Laing, Carmody, and Peck 1976a).

The genetic basis of Barr's subspecies of *N. tellkampfi* can be examined more closely. Of the four subspecies, *N. t. meridionalis* is the most distinct morphologically, while *N. t. henroti* appears most distinct genetically because of the absence of hybrids with other subspecies (Barr 1979). Nei's index makes no particular sense with either of these criteria (Table 5–8). *N. t. henroti* is very similar to *N. t. tellkampfi* ($\bar{I}$ = 0.970) from which it probably arose (Barr 1979). *N. t. viator* is the most distinct genetically, even though it forms hybrids with *N. t. tellkampfi* and *N. t. meridionalis*.

Genetic Differentiation in Aquatic Populations

There are no studies of the scope and magnitude of Peck's and Kane's for cave-limited aquatic species, but the work of Dickson et al. (1979) on the amphipod *Crangonyx antennatus* shows some potentially interesting contrasts and similarities with terrestrial species. They analyzed six populations in an area of about 50 km^2 in southwestern Virginia (Fig. 5–2). Since troglobitic aquatic species are often found in other interstitial habitats, such as wells, populations should be less isolated

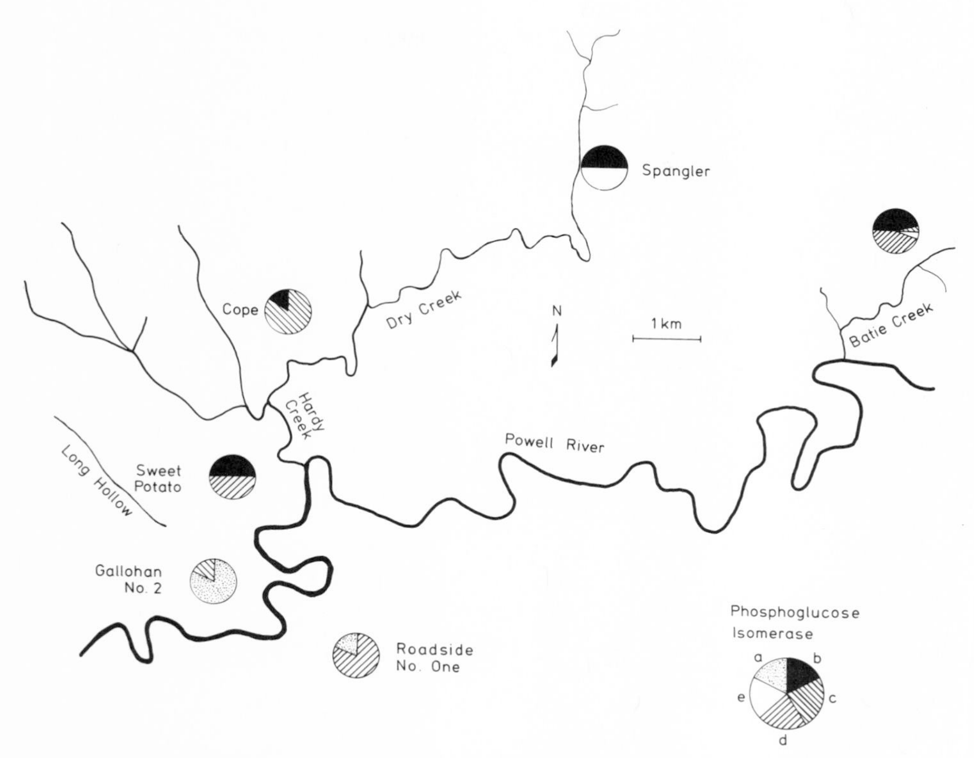

Figure 5–2 Distribution of five alleles of phosphoglucose isomerase in the amphipod *Crangonyx antennatus* from caves in Lee County, Virginia. (From Dickson et al. 1979.)

than terrestrial populations (Culver, Holsinger, and Baroody 1973). However, even though Dickson and associates looked at a smaller area than the range of *N. t. tellkampfi,* they found population differentiation intermediate between that of *N. t. tellkampfi* and *P. hirtus* (see Table 5–1). The average Nei index for *Crangonyx antennatus* was 0.920 and ranged from 0.858 and 0.968 (Dickson et al. 1979). For six of the eight loci studied, the same allele (or at least the same electromorph) was fixed in all six populations. The most distinct population was the one from Roadside Cave No. 1, which is separated from the other populations studied by the Powell River.

The most interesting pattern of variation is shown by the phosphoglucose isomerase. Not only is it clear that the Roadside Cave No. 1 population is the most distinct, this locus provides the only evidence of genetic differences between pool and stream populations. Dickson (1977) found numerous morphological differences. One allele of phos-

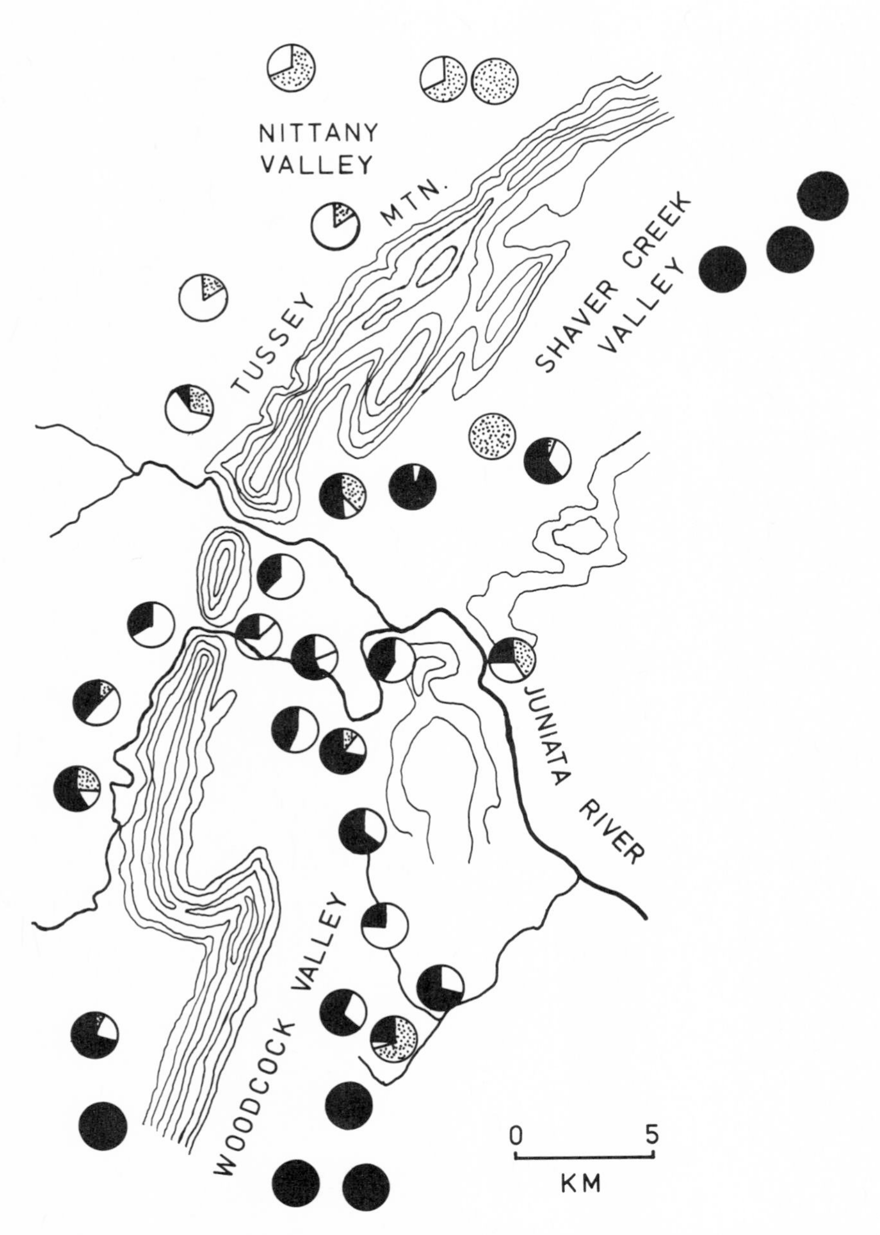

Figure 5–3 Distribution of three alleles of malate dehydrogenase in the amphipod *Gammarus minus* from springs in Pennsylvania. (From Gooch and Golladay 1981.)

phoglucose isomerase was found in all pool populations and in no stream populations. Since only five individuals in each population were scored for phosphoglucose isomerase, there is a distinct possibility that some stream populations may have the allele. However, its high frequency in pool populations ($\bar{p}$ = 0.57) indicates some degree of difference between stream and pool populations.

One possible explanation of the unexpectedly high degree of differentiation of *Crangonyx antennatus* populations lies in the pattern of differentiation of their surface ancestors. Possible analogs of the surface ancestors of *C. antennatus* are populations of the amphipod *Gammarus minus* in Pennsylvania springs, which show considerable differentiation at the malate dehydrogenase locus (Fig. 5–3). Gooch and Golladay (1981) suggest that the one-dimensional nature of the habitat may be the cause.

Summary and Prospect

It should be abundantly clear that no easy generalizations about the genetic variability of cave populations are possible. Certainly not all cave populations have low heterozygosity, as the simplest forms of both neutral mutation theory and selection theory predict. Comparisons with epigean populations are difficult not only because of problems of determining closely related surface ancestors but also because of difficulties in extracting all variation at a locus, the small number of loci studied, and the fact that different loci have been studied by different investigators.

A more modest, but potentially more fruitful, way to proceed is to study differences among cave populations. For example, there is more genetic than morphological differentiation for most species studied. The important advantage of these studies is that the same enzyme systems are compared. If it becomes possible to compare the same enzyme systems for very different species, for example, millipeds and beetles, considerable progress may come out of the present confusion. In particular, two hypotheses could be tested. The first is Barr's (1968) conjecture that cave adaptation involves considerable reorganization of the genotype. There is no evidence from available data that trogloxenes and troglophiles have reduced genetic variability as a result of this reorganization, but species in the same phyletic line need to be compared. The second is Kane's (1981) conjecture about the importance of ecological differences in explaining differences in heterozygosity, which could be tested by comparing the same enzyme systems.

Species Interactions and Community Structure

Questions concerning the nature and importance of species interactions have become increasingly controversial. In the mid-1960s it was thought that interspecific competition was the major organizing factor in many communities. Through the pioneering work of the late Robert MacArthur and his students, many aspects of the structure and dynamics of bird communities seemed best explained by competitive interactions, which could be described in a general way by the Lotka-Volterra competition equations. This work, or at least its generality, has been challenged on three levels. First, there is the question of how important species interactions are in general and whether the interactions are competitive, parasitic, predatory, or mutualistic. For example, Connell (1975) noted that a significant fraction of, but by no means all, communities are physically rather than biotically controlled. In addition, there has been a tendency in the last few years to concentrate on single-species demography. While interactions between species can still be considered in the guise of age-specific fecundity and mortality effects, the emphasis is away from a coevolutionary perspective and toward a individualistic concept of communities.

Second, the question of which type of interaction is most important in structuring a particular community is very much open. Many ecologists have pointed out that the birds studied by MacArthur and his students were at or near the top of the food chain. There is no *a priori* reason to expect that competition will also be important lower in the food chain. Thus Connell argues that predation is a more widespread and important interaction. Furthermore, some of the best examples of competition, such as character displacement in the Galápagos avifauna, are probably not examples of competition at all (Strong, Szyska, and Simberloff 1979). Until very recently the debate revolved around the relative importance of competition and predation. However, Risch and Boucher (1976) and Price (1980) have forcefully argued that whole classes of interactions have been ignored. Risch and Boucher, among others, have argued that mutualistic interactions may be the key to understanding community structure, while Price argues the same for parasitism.

Third, there has been widespread disillusionment with the utility of simple mathematical models to describe species interactions. In part, this disillusionment stems from the very real difficulties in measuring interaction coefficients and from the failure of almost all suggested shortcuts to measuring interaction coefficients. But in part it comes from the failure of the models to predict observed patterns (Neill 1974).

These three questions will be considered throughout this chapter in various ways. The first section reviews what is known about the relative importance of different interactions in cave communities. The second section considers in some depth the beetle–cricket egg interaction, and the third section reviews in detail the interactions in cave stream communities in the southern Appalachians, the most thoroughly studied case. The concluding section suggests that most questions about species interactions are ill posed and argues for the central importance of models.

Which Interactions Are Important?

As Price (1980) points out, the role of parasites in the structure and dynamics of most communities has been greatly underestimated. Even the inventory of parasites of cave organisms is very incomplete (Vandel 1964), and there are almost no data on the percentage of the host population infected or the effect of the parasite on the host. Even the meager data available suggest that parasites may be important in some communities. Keith (1975) found that almost all *Pseudanophthalmus tenuis* beetles in Murray Spring Cave in Indiana were infected with the fungal parasite *Laboulbenia subterranea,* with an average of up to fifteen infestations per beetle. The effect on the beetles of these symbionts on the integument is not known, and in fact they may be commensals rather than parasites. Extremes of specialization, even for parasites, occur in caves. One of the most spectacular examples are the Temnocephala, parasitic platyhelminth worms intermediate in morphology between turbellarians and trematodes, which parasitize European cave shrimp. Matjašič (1958) reported that seven species and several genera of Temnocephala are found only on the cave shrimp *Trogocaris schmidti,* with each species specializing on a particular region of the body. For example, *Subtelsonia perianalis* is found around the anus of *T. schmidti.*

A similar level of ignorance obtains for mutualistic interactions. No free-living mutualists have been reported from caves, but ectosymbionts that are probably mutualistic or commensal are known. Hobbs (1973, 1975) studied the entocytherid ostracods that live on the exoskeletons of cave crayfish. They feed on microorganisms and on detritus that accumulates on the host exoskeleton and are unable to complete their life cycle away from their host. Crayfish probably derive some advantage from the cleaning activity of the ostracods, but the main effect of the interaction is benefit to the ostracods. Hobbs compared the ostracod symbionts of the cave-limited *Orconectes inermis* to those of the facultative cave dweller *Cambarus laevis*. Almost all the ostracods on *O. inermis* were *Sagittocythere barri,* which is rarely found on other species in Hobbs' study area. In contrast, *C. laevis* commonly harbored three species. Infestation and reinfestation occur when the hosts copulate, when ostracod eggs become attached to newly hatched crayfish carried under the abdomen and, following a molt, when the exuviae are eaten by the crayfish. Levels of infestation are lower in the cave-limited species, but this is complicated by the strong effect of crayfish size on the number of ostracod infestations (Table 6–1). Since the cave-limited species are smaller, fewer ostracods per crayfish are expected. Regression analysis indicates that a larger minimum size is required for infestation of the cave-limited species than of the facultative cave dwellers, but that the rate of infestation increases more rapidly with size in troglobitic species. This may be a consequence of the specialization of *S. barri* on *Orconectes inermis,* but the adaptive significance, if any, is not clear.

Christiansen and Bullion (1978) attempted to assess the importance of competition and predation for terrestrial cave fauna in the Haute-Garonne and Ariège regions of France. Their basic procedure was to visually census for 100 minutes the populations of about fifty terrestrial

Table 6–1 Numbers of entocytherid ostracods inhabiting various crayfish, and the relationship ($y = a + bx$) between crayfish carapace length (x) and number of ostracod infestations (y); b is the slope and a is the intercept of the regression equation. Minimum size is the smallest crayfish expected to harbor ostracods, based on the regression analysis. (Data from Hobbs 1973, 1975.)

Species	Habitat	Ecological range	$\bar{y}$ (± 95% confidence interval)	a	b	Minimum size (mm)
Orconectes inermis inermis	Cave	Cave-limited	18.8 ± 3.6	−28	1.95	15
Orconectes inermis testii	Cave	Cave-limited	20.2 ± 13.6	−45	2.80	16
Cambarus laevis	Cave	Facultative cave-dweller	26.5 ± 3.7	6.2	0.60	0
Cambarus bartonii	Surface	Occasionally in caves	119 ± 17.5	—	—	—

species and to estimate various environmental parameters, such as dry speleothems, calcareous clay, guano, and standing crop of organic debris, for fifty-eight caves. Some classification of caves was made, distinguishing underground rivers, vertical sinks, and so on, as well as the aphotic and entrance zones. Each environmental parameter was rated on a scale of 1 to 6. They then attempted to determine what affected the abundance of various species by stepwise multiple regression techniques. Their study was almost exclusively a between-cave comparison and did not detect all interactions, such as competition resulting in microhabitat separation within a cave. Christiansen and Bullion themselves pointed out some statistical weaknesses and noted that some climatic variables were not measured, but their study did provide insights into the importance of species interactions. There were examples of apparent competitive exclusion between troglobitic omnivore and troglophilic carnivore beetles but not among troglobitic carnivores or troglophilic omnivores. That is, competitive exclusion occurred but was infrequent.

Collembola species were analyzed more completely. Table 6–2 summarizes the results for three cave-limited Collembola: *Tomocerus problematicus*, *Pseudosinella theodoridesi*, and *P. virei*. Most of the major negative correlates are other species rather than physical factors, and most are probably competitors rather than predators. Possible competitors include other Collembola, millipeds, and bathyscine

Table 6–2 Summary of major factors affecting the abundance of *Tomocerus problematicus*, *Pseudosinella theodoridesi*, and *P. virei* in caves in southern France. (Data from Christiansen and Bullion 1978.)

	Correlates		
Rank	*T. problematicus*	*P. theodoridesi*	*P. virei*
Negative			
1	Other Entomobryidae[1]	Other Entomobryidae[1]	*P. superduodecima*
2	*T. minor*	Dry speleothems	Cave length
3	Millipedes	*P. impediens*	Carabid beetles
4	*P. impediens*	Bathyscine beetles	*P. impediens*
5	*P. virei*	*T. minor*	Mites
Positive			
1	Noncalcareous clay	Altitude	Guano
2	Diplurans	Organic debris	Breakdown
3	Calcareous clay	Diplurans	Organic debris
4	Altitude	Calcareous clay	Wet speleothems
5	Standing water	Opilionids	Sand and silt

1. *Pseudosinella sexoculata*, *P. alba*, *Heteromurus nitidus*, and *Lepidocyrtus* spp.

beetles. The only certain predator effect listed is that of carabid beetles on *Pseudosinella virei*. Therefore, predation appears to be much less important than competition in determining community structure. Among the positive correlates listed in Table 6–2, almost all are environmental and resource parameters, but Diplura are positively correlated with both *T. problematicus* and *C. theodoridesi*.

One correlation not included in the table is a very strong positive correlation between *T. problematicus* and *P. theodoridesi*. Christiansen and Bullion felt that this indicated joint correlation with other variables rather than a mutualistic interaction, but they did not attempt to confirm that. Instead, they deleted the abundance of one species in the stepwise regression analysis of the other species. The authors may be correct in assuming that the correlation between *T. problematicus* and *P. theodoridesi* is spurious, but they also share the bias of most ecologists, at least until very recently, that mutualisms outside the tropics are rare. It is at least possible that the two species are mutualists. In the laboratory, successful establishment of culture jars is often facilitated by the presence of a reproducing population of another species (Culver 1974), indicating that mutualistic effects do occur.

In contrast to Christiansen and Bullion's study, Kane's (1974) study of terrestrial cave communities in Mammoth Cave National Park gives some hints that predation may be much more important than competition. To attract organisms, Kane set out leaf litter in m^2 quadrats. In Little Beauty Cave, between 36 and 40 percent of the species attracted, depending on the quadrat, were predators, and between 9 and 15 percent of the individuals were predators. In the Natural Bridge area of Mammoth Cave, between 38 and 39 percent of the species and between 8 and 48 percent of the individuals were predators. The large and relatively constant ratio of predator species to prey species is, at the least, consistent with Kane's hypothesis that predation largely controls community structure.

There is also evidence for competition among predators. Van Zant, Poulson, and Kane (1978) claimed that character displacement occurred when two small beetle predators, *Pseudanophthalmus menetriesii* and *P. pubescens* occurred together. In caves where both were found, *P. pubescens* was between 4.7 and 4.8 mm long, and *P. menetriesii* was between 4.4 and 4.5 mm long. In the one cave where only *P. menetriesii* was present, it was 4.6 mm long. The authors speculate that this difference is due to differences in sizes of prey taken. Both species feed on small invertebrates such as Collembola. Barr and Crowley (1981), however, suggest that the size differences of *P. menetriesii* are clinal and unrelated to competition.

There are no similar studies of aquatic cave communities available, but a few comments can be made. In several major cave regions, such as the Appalachians, detritivores such as isopods and amphipods are at the top of the food chain. The question of whether competition or predation is more important among the macroscopic fauna is often trivial because this fauna has no predators. The absence of large predators in a particular area is most likely caused by historical factors (see chapter 7). In many caves with fish predators, macroscopic detritivores are very rare or absent because the streams are mud-bottomed, a generally unfavorable amphipod and isopod habitat. One example of an aquatic community with macroscopic predators and competitors will be considered in the section on Appalachian cave stream communities.

The Beetle–Cricket Interaction

One terrestrial predator-prey relationship that has received particular attention is the interaction between carabids and cave cricket eggs (Fig. 6–1). Over 75 percent of the diet of *N. tellkampfi* is eggs and

Figure 6–1 *Rhadine subterranea* eating cricket egg. (Photo courtesy of Dr. Robert W. Mitchell, Department of Biology, Texas Tech University, Lubbock, Texas.)

nymphs of *H. subterraneus* (Norton, Kane, and Poulson 1975), with eggs the preferred food (Kane and Poulson 1976). Like most other cave crickets, *Hadenoecus subterraneus* is an omnivore and obtains nearly all its food outside the cave. Although no long-term studies have been done, there can be little doubt that this predator–prey pair have a major effect on each other's population sizes. In an ingenious experiment in which *N. tellkampfi* beetles were excluded by a low barrier that did not prevent the crickets from ovipositing, Kane and Poulson (1976) found that *Hadenoecus* egg densities in beetle-free enclosures were about ten times higher than in the surrounding area. *N. tellkampfi* ate between 72 percent and 97 percent of the eggs oviposited in the surrounding area.

Hubbell and Norton (1978) found one morphological difference in preyed-upon and non-preyed-upon populations: ovipositor lengths were significantly longer in preyed-upon populations. Apparently eggs buried deeper in the sand are more difficult for beetles to locate. The crickets show a peak in egg laying in early spring that coincides with or slightly precedes the resumption of epigean feeding (Hubbell and

Norton 1978). This seasonality of egg laying produces a seasonality in the life cycle of the beetle, with a sharp increase in the emergence of teneral *Neaphaenops* about three months after the peak of cricket egg laying (Fig. 6–2).

Because these beetles are almost certainly the major cause of cricket mortality, and crickets are the major source of food for the beetles, the dynamics of the interaction are particularly interesting. It is a sad commentary on the gap between theoretical and field ecology that in spite of extensive work on this interaction, there are no data available to make any but the most general application of predator–prey models. The following paragraphs are speculations and suggestions for a closer connection between theory and field work.

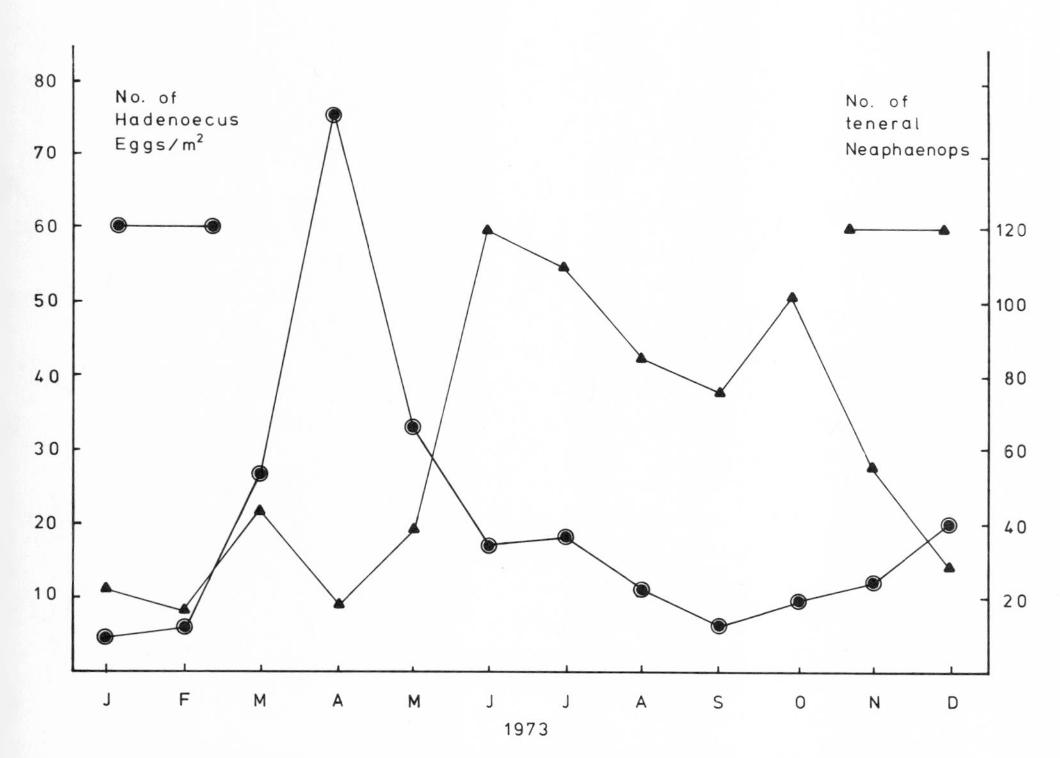

Figure 6–2 Seasonal changes in number of eggs per m² of the cave cricket *Hadenoecus subterraneus* and a visual census of newly emerging adults (tenerals) of the beetle *Neaphaenops tellkampfi* in Edwards Avenue, Great Onyx Cave, Kentucky. (Date from Kane, Norton, and Poulson 1975 and Norton, Kane, and Poulson 1975.)

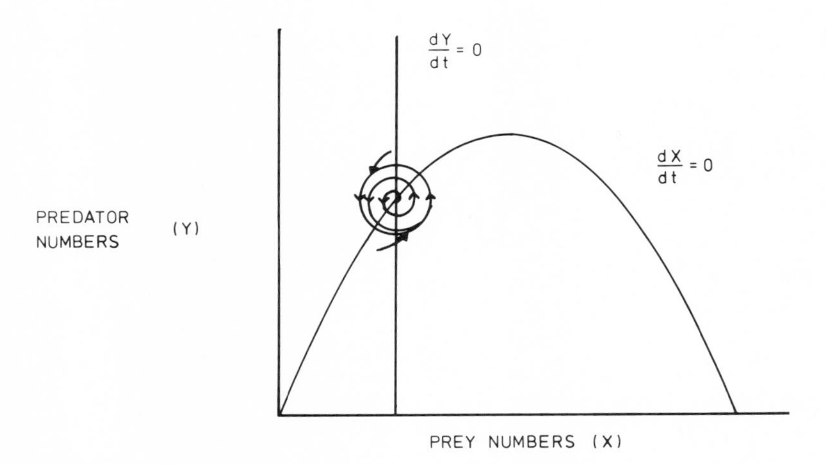

Figure 6–3 Conjectured dynamics of the beetle–cricket egg interaction, based on Rosenzweig and MacArthur's (1963) predator–prey model. The circle is a stable limit cycle, resulting in population oscillations.

One of the simplest predator–prey models is the graphical model of Rosenzweig and MacArthur (1963). The stability of the equilibrium is determined by finding whether the predator isocline passes through the prey isocline to the right or left of the hump (Fig. 6–3), with damped oscillations occurring to the right and limit cycles or extinction to the left (May 1972). The predator isocline is determined by the ratio of predator mortality to predation rate. Since predation rates on cricket eggs are very high, thus moving the predator isocline to the left, it is quite possible that the intersection is to the left of the hump of the prey isocline, resulting in a stable oscillation (May 1972).

A more realistic model would include both the time lags in the system due to development time of the beetles and seasonal availability of cricket eggs. One interesting comparison would be between the *Neaphaenops* and *Hadenoecus* in Kentucky, where eggs are present only in spring and summer, and the beetle *Rhadine subterranea* and cave crickets in the genus *Ceuthophilus* in Texas caves. There, cricket eggs are available throughout the year because both a summer egg-laying and a winter egg-laying species are present—*C. cunicularis* and *C. secretus* (Mitchell 1968). One would expect greater seasonal and long-term fluctuations in population sizes of both species in the Kentucky system.

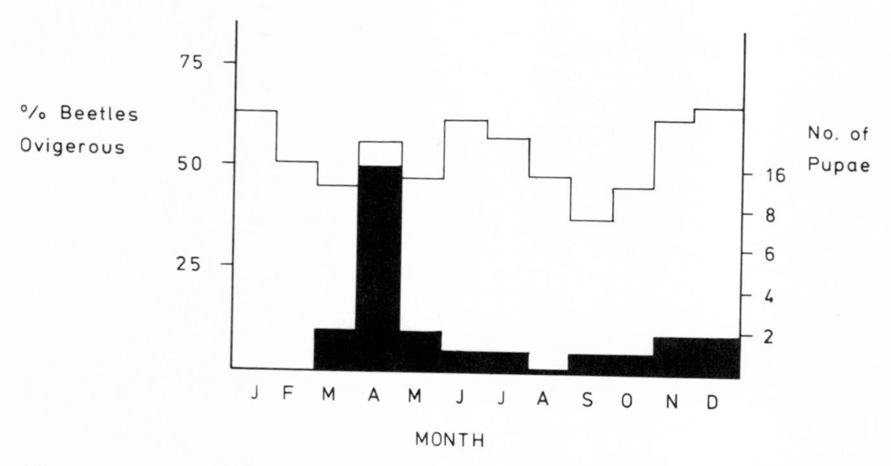

Figure 6–4 Monthly percentages of ovigerous *Neaphaenops tellkampfi* (unshaded bars) and total numbers of pupae (shaded bars) in Edwards Avenue, Great Onyx Cave, Kentucky, in 1973. (Data from Kane, Norton, and Poulson 1975.)

Time lags themselves are of special interest. Although they are generally destabilizing, the form of the lag is very important (MacDonald 1978). If x is population size and z is some lag function

$$\frac{dx}{dt} = f(x, z) \tag{6-1}$$

then the case of constant lag time, T, in other words, $z = x(t - T)$, is less stable then when lag times vary, as:

$$z = \int_{-\infty}^{t} x(\tau)G(t - \tau)\, d\tau \tag{6-2}$$

where τ is a varying time lag. It is likely that the lag function G (more properly the memory function) varies for different life history stages of *N. tellkampfi*. Most striking are the differences between monthly percentages of ovigerous beetles, which are mostly uniform throughout the year, and number of pupae, which show a sharp increase in the spring (Fig. 6–4).

Appalachian Stream Communities

Species interactions in aquatic cave communities have been studied most extensively in the caves of the southern Appalachians, particularly in three areas: the Monongahela River Valley and the Greenbrier River Valley of West Virginia and the Powell River Valley of Virginia and Tennessee. A brief description of the fauna and its physical setting is necessary to understand the interactions that occur.

In almost all caves, the stream fauna is dominated by amphipods and isopods. In the Powell River Valley the isopods *Caecidotea recurvata* (Fig. 6–5) and *Lirceus usdagalun* and the amphipod *Crangonyx antennatus* predominate; in the Greenbrier River Valley *Caecidotea holsingeri* and the amphipods *Gammarus minus, Stygobromus emarginatus,* and *Stygobromus spinatus* predominate; and in the Monongahela River Valley *Caecidotea cannula* and *C. holsingeri* predominate. There are few other macroscopic detritivores in the stream. Very occasionally, other amphipod and isopod species, for example, *Caecidotea richardsonae,* are in the Powell River Valley, but these almost always replace another species, in this case *C. recurvata.* In a few Greenbrier River Valley caves, the crayfish *Cambarus nerterius* is common, but these caves have large, mud-bottomed streams instead of the small, gravel-bottomed streams where amphipods and isopods occur. Snails in the genus *Fontigens* occur sporadically in all three drainages.

Considerable information is available about the evolutionary history of the major amphipod and isopod genera. In a recent revision of *Stygobromus,* Holsinger makes a strong case that there was a freshwater invasion from marine waters during the late Paleozoic or early Mesozoic and that the subsequent invasion of caves occurred from interstitial rather than epigean habitats. Although the evolutionary history of *Caecidotea* is less well studied, and their nomenclatural history is intricate, there are parallels with *Stygobromus* (Steeves 1969) that suggest a similar history, if not so ancient. *Crangonyx* is a more recent cave inhabitant than *Stygobromus,* although how recent is unknown (Holsinger 1969). The genus is also found in streams and springs and may have invaded caves directly from streams or via interstitial habitats. *Lirceus usdagalun* probably invaded caves from springs or streams, but its relationships to other *Lirceus* species are obscure (Holsinger and Bowman 1973). Finally, *Gammarus minus* is common in springs and spring runs as well as caves, so it is clearly a recent cave invader from surface waters.

Figure 6–5 The isopod *Caecidotea recurvata*. (Photograph by author.)

Aquatic predators are generally uncommon. Planarians are locally common but probably feed mostly on injured or moribund individuals. The only predators of note are larvae of the salamander *Gyrinophilus porphyriticus,* which are especially important in many Powell River Valley caves.

In common with gravel-bottom streams on the surface, nearly all cave streams in the southern Appalachians alternate between deeps (pools) and shallows (riffles). The riffles are much shorter than the pools and repeat at a more or less regular interval of five to seven stream widths (Leopold, Wolman, and Miller 1964). The formation and maintenance of riffles is a fascinating topic in its own right, but there are two points of biological interest. First, larger rocks lie on top of smaller rocks, and second, individual rocks move from riffle to riffle, especially during spring floods, but the position of riffles stays the same.

The majority of amphipods and isopods are found in riffles rather than in pools. For example, in Benedict's Cave in Greenbrier County, West Virginia, the population density in riffles was more than five times that of pools. There are several explanations for the greater densities in riffles. First, there is more dissolved oxygen in the water; second, riffles act as detritus traps and so more food is available; and third, salamander predators, when present, are concentrated in pools, where their lateral line system functions more efficiently for detecting distant prey.

The riffles themselves present problems to isopods and amphipods. The evolutionary history of many of the species has been in slow-moving interstitial water, and they are especially vulnerable to currents. Many individuals cannot maintain their position in the current of a cave stream even if they are clinging to the top of a rock. The major exception to this vulnerability seems to be *Lirceus usdagalun,* which is the only species that is at all common on the tops of rocks, and then only in very slow-moving streams. In the absence of predators, dislodgement is the major source of mortality. Dislodged animals frequently suffer appendage damage in laboratory streams, and many amphipods and isopods collected in natural streams have appendage damage. Since the gravels themselves move, at least during floods, immediate mortality is important as well. The field evidence is consistent with this. In Benedict's Cave, for example, the low point of the *Gammarus minus* population corresponded to early spring flooding, and population size did not reach preflood levels for four months, indicating a real population drop rather than a movement into areas inaccessible to sampling (Culver 1971a).

The Basis of Competition In most of the caves, amphipods and isopods use the undersides of rocks and gravels primarily as refuges from the brunt of the current, but they are also places to feed and to hide from salamanders. Although there are almost always many more rocks than there are animals, the percentage of amphipods washed out by the current, at least in laboratory streams, is density dependent, indicating that competition is a factor (Fig. 6–6). The most parsimonious explanation of this is that washouts are primarily the result of encounters between individuals. The volume of water on the underside of

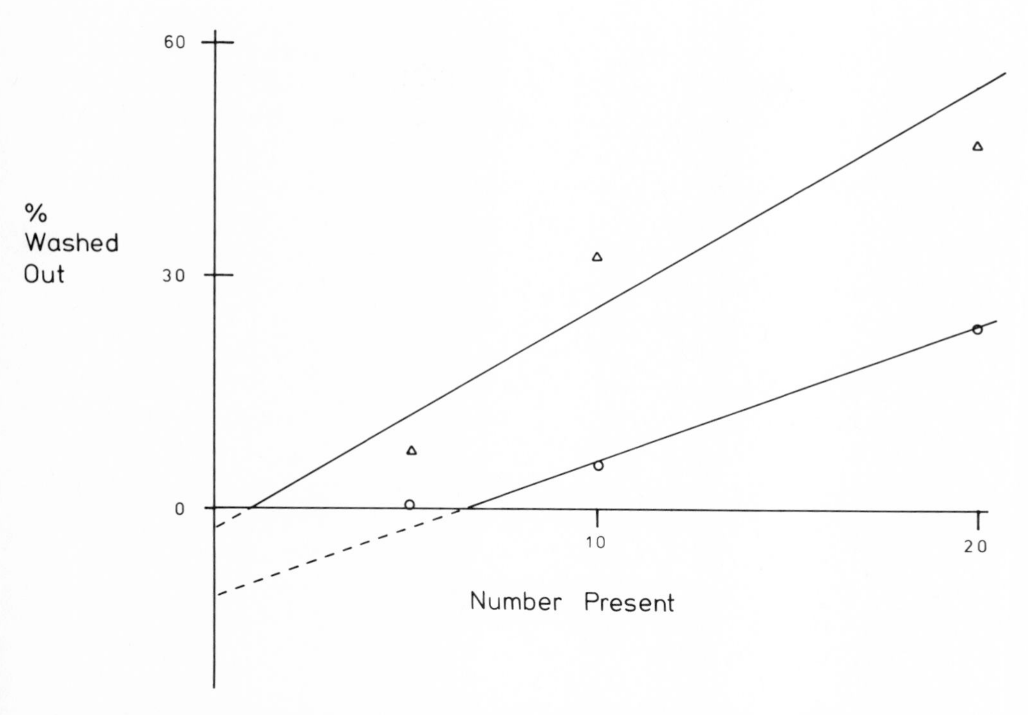

Figure 6–6 Percentage of *Gammarus minus* washed out of an artificial stream in 24 hours in relation to the number originally present. In the experiment represented by the upper curve, paper "detritus" was present, and in the experiment shown by the lower curve, leaf detritus was present. The rise in both curves indicates density dependence. (From Culver 1971a.)

a rock out of the brunt of the current is actually quite small (Ambühl 1959), and there is considerable movement or dislodgement of animals among rocks in the same riffle. The hypothesis of density dependence of the washout rate is augmented by the avoidance behavior displayed by most individuals toward others of the same and of different species (Culver 1970a). For example, in experiments in a mud-bottomed finger bowl with one rock, a single *Caecidotea holsingeri* strongly preferred the rock when alone, but was excluded from it when either *Stygobromus emarginatus* or *Gammarus minus* was present (Culver 1970a).

The studies described above have identified a mechanism of competition, but the question remains of how important it is in nature. There have been two somewhat overlapping approaches to studying competition. The first is to carefully document that competition is occurring and to design experiments that can directly falsify the hypothesis that the community is competitively controlled. One of the best examples is the work of Reynoldson and his colleagues on triclad flatworms (Reynoldson and Bellamy 1971). The strength of this approach is clearly the testing of a falsifiable hypothesis. Its weakness is that the hypothesis being tested may not be what one thinks it is. For example, the prediction that the abundance of competing species should be negatively correlated through time does not test the hypothesis that species are competing, but rather that the competition is of a particular kind. This example will be elaborated later in the chapter.

The second approach is not to test for competition directly but rather to explore its consequences, usually with the aid of models. The work of Diamond (1975) exemplifies this approach. Its strength is that a relatively large number of predictions can be made, and its weakness is the danger of being right for the wrong reasons, as Paul Dayton has aptly put it. This approach can result in a castle built on sand; the first approach can result in a strong foundation with no castle. The arguments developed below attempt to use the strengths of both approaches. The most general consequences of competition will be considered first, followed by predictions that depend on the kind and intensity of competition.

Niche Separation One of the most universal results of competition is niche separation. Particularly interesting are populations that show a niche difference in allopatry and sympatry. A qualitative view of niche shifts in the Greenbrier Valley stream fauna is shown in Figure 6–7. Two species, *Gammarus minus* and *Stygobromus spinatus,* do not show any significant niche shift when in the same cave stream (syntopy). *Stygobromus spinatus* is found deep in riffles and does not usually encounter any other species, so it is not surprising that it does not undergo any niche shift. *Gammarus minus* occurs near the top of riffles, where it overlaps with *Stygobromus emarginatus* and *Caecidotea holsingeri*. In syntopy with *G. minus,* these two species are excluded from riffles. *Caecidotea holsingeri* is limited to pools, and *S. emarginatus* is limited to tiny trickles of water feeding into the stream. All four species feed on dead leaves and their microflora, and there is no evidence of shifts in food eaten when in syntopy (Culver 1970a).

Estes (1978) has made a more detailed study of the microhabitat niche of *Lirceus usdagalun* in two caves in the Powell Valley. In Gallohan Cave No. 1, there are significant populations of *Caecidotea recurvata* and *Crangonyx antennatus,* in addition to *L. usdagalun*. In the area of Thompson Cedar Cave sampled by Estes, *C. recurvatus* and *C.*

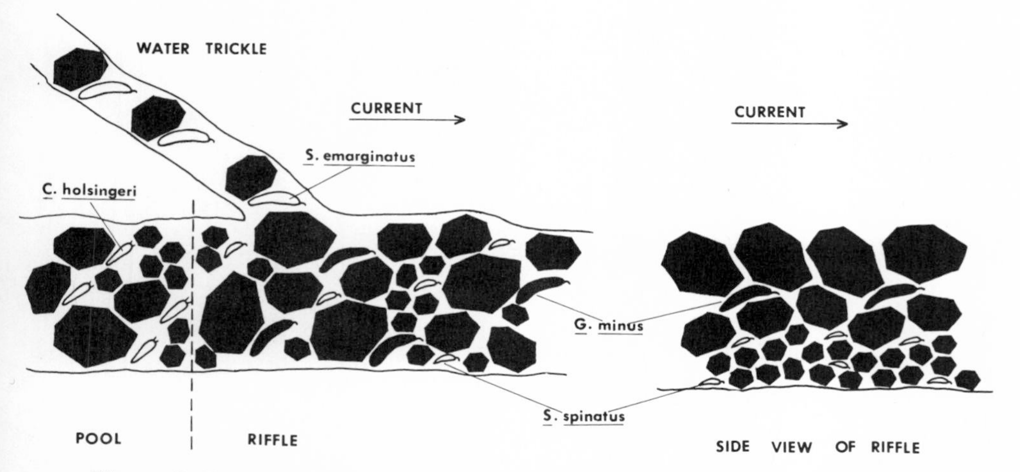

Figure 6–7 Diagrammatic view of niche separation of *Caecidotea holsingeri* (open oblong shapes), *Gammarus minus* (large solid crescents), *Stygobromus emarginatus* (large open crescents), and *Stygobromus spinatus* (small open crescents) in caves of Greenbrier County, West Virginia. *C. holsingeri* and *S. emarginatus* undergo niche shifts in the presence of competitors. No more than three of the species actually occur together in a single cave stream. (Drawing by Christine Turnbull.)

antennatus are almost completely absent, but *L. usdagalun* is common. Estes sampled six microhabitats and two velocity:depth profiles (with 0.67 used as an arbitrary dividing line between the two). Estes compared densities for each microhabitat at each velocity:depth profile and found a significant reduction in the density of *L. usdagalun* in Gallohan Cave No. 1 on bedrock in slow current, among small rocks in slow current, and among gravels in slow current. This difference is almost certainly due to the presence of competitors in these three areas. The niche breadth (B) of *L. usdagalun,* calculated using

$$B = 1\Big/\sum_i p_i^2 \qquad (6\text{-}3)$$

where p_i is its frequency in microhabitat-velocity type i, is greater in the absence of competitors (B = 7.8) than in the presence of competitors (B = 5.7), as expected. It is clear that current velocity plays a major role in niche separation. With each microhabitat–velocity:depth type weighted equally to facilitate comparison, there is a shift of *L. usdagalun* toward faster currents when competitors are present. In the

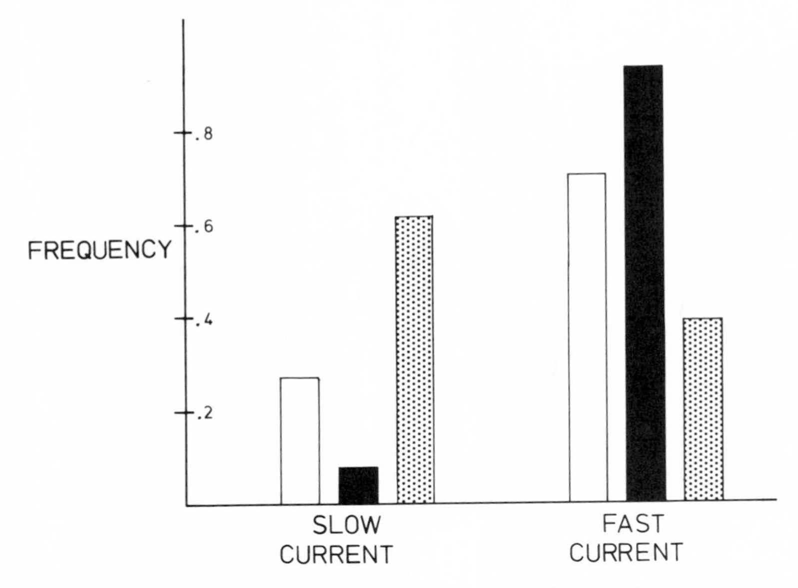

Figure 6–8 Relative frequencies of *Lirceus usdagalun* and its competitors in microhabitats of slow and fast current. The open bars represent the frequencies of *L. usdagalun* in the absence of competitors; the solid bars are the frequencies of *L. usdagalun* with competitors present; the shaded bars are frequencies of competitors (*Caecidotea recurvata* and *Crangonyx antennatus*). (Data modified from Estes 1978.)

absence of competitors, approximately 70 percent of the population occurs in fast currents, but over 90 percent occurs in fast currents when competitors are present (Fig. 6–8). Estes suggests that the success of *L. usdagalun* in Thompson Cedar Cave is due at least in part to the higher current velocities in that cave.

Population Size Changes Due to Competition Simply stated, the summed abundances of competing species should vary less through time than the abundances of any individual species. This prediction makes general sense where the total food or habitat available remains constant, and the prediction can be developed more formally as follows (Culver 1981). Consider two competitors whose abundances are N_1 and N_2:

$$\frac{dN_1}{dt} = f_1(N_1, N_2), \frac{dN_2}{dt} = f_2(N_1, N_2) \qquad (6\text{-}4)$$

Let $n_1 = N_1 - \hat{N}_1$ and $n_2 = N_2 - \hat{N}_2$, where $\hat{N}_i$ are the equilibrium populations. Then

$$\frac{dn_1}{dt} = \frac{\partial f_1(\hat{N}_1, \hat{N}_2)}{\partial N_1} n_1 + \frac{\partial f_1(N_1, N_2)}{\partial N_2} n_2 + \text{higher-order terms}$$

$$\frac{dn_2}{dt} = \frac{\partial f_2(\hat{N}_1, \hat{N}_2)}{\partial N_1} n_1 + \frac{\partial f_2(N_1, N_2)}{\partial N_2} n_2 + \text{higher-order terms} \tag{6-5}$$

Near equilibrium the higher-order terms vanish. If competition is occurring, the partial derivatives in parentheses are negative. Using a convenient shorthand, the above equations can be simplified and extended to n species:

$$\dot{n} = An \tag{6-6}$$

where $\dot{n}$ is a column vector with elements dn_i/dt and A is a matrix of partial derivatives as in equation 6-5.

$$Av = \lambda v \tag{6-7}$$

where λ is a scalar eigenvalue and v is the associated eigenvector. (For an explanation of eigenvalues and eigenvectors in an ecological context, see Roughgarden 1979.) The real parts of the eigenvalues measure the community's rate of return to (or departure from) equilibrium following a perturbation. Associated with each eigenvalue is an eigenvector, which is a linear combination of the deviations of population sizes from equilibrium. The rate of return following a particular perturbation depends on the magnitude of the eigenvalue associated with the eigenvector that most closely approximates the perturbation. The Perron-Frobenius Theorem (Gantmacher 1959) demonstrates that the eigenvector associated with the largest eigenvalue has all positive elements. A perturbation that changes the abundance of all species in the same direction will damp out more quickly than any other perturbation. Thus total abundance should vary less through time than the abundance of an individual species.

This competition hypothesis can be tested by comparing the variance of total abundance, which is controlled by the largest eigenvalue, to the sum of the individual species variances, which are controlled by smaller eigenvalues. In three of the four caves studied in the Powell Valley, the variance of total abundance is less than the sum of the variances of abundances of individual species (Table 6–3). The variance ratios range from 2.39 in Gallohan Cave No. 2 to 0.98 in Gallohan No. 1. None are statistically significant. If these variance ratios are typical for communities of competitors, there will rarely be sufficient data to demonstrate statistical significance.

All in all, the fit of the data to the prediction is poor. In part this may be because of small sample sizes, but it also seems likely that some other process is involved. Either competition is not occurring, or the null hypothesis of no correlation in the absence of competition is incorrect. The hypothesis above was generated for the situation in which equilibrium is fixed and the population sizes are subject to random perturbation about this equilibrium. But the equilibrium itself may vary because of changes in current or food availability. If carrying capacities (K's of the standard competition equations) of species are positively correlated through time, then the variance in total abundance may exceed the sum of the variances of individual species' abundances. For example, if the carrying capacities of two competitors are both low in winter and high in summer, then species abundances might be positively correlated even though they are competing.

Correlation of abundances of pairs of species through time provides much stronger evidence for competition, but the predictions depend on the intensity of competition, so discussion of these data will be deferred until after the measurement of competition is discussed. It should be noted in passing that all *partial* correlations of competitor abundances should be negative (if carrying capacities are not too strongly correlated), and eight of ten such partial correlations are negative for the amphipod-isopod systems, three significantly so.

Table 6–3 The ratio of the sum of the variances of individual species abundances (ΣV_i) to the variance of total abundance (V_T). If species are competing, this ratio should be greater than unity. The number of samples required for statistical significance ($p > 0.95$), if the ratio observed is the true variance ratio, is listed in the column labeled N*. (Data from Culver 1981.)

Cave	No. species	No. time samples	$\Sigma V_i / V_T$	N*
Gallohan No. 2	3	4	2.39	12
Court Street	3	7	1.77	25
Spangler	2	4	1.12	> 100
Gallohan No. 1	3	5	0.98	> 100

Measurement of Competition Coefficients Any model of competition among these amphipods and isopods must take into account the physical structure of the stream. A riffle is a patchy environment from the point of view of an amphipod or isopod. It consists of a set of habitable patches (the undersides of rocks) separated by uninhabitable areas where the animals face the brunt of the current. Competition is for places to avoid the brunt of the current (or for feeding sites), and such places are in short supply whenever two individuals meet. The continued movement or dislodgement of animals among rocks in a riffle results in competition even when both species are rare.

This view of a riffle as a series of tiny islands is supported not only by the laboratory stream studies mentioned earlier but also by field observations. In such a system one would expect to see turnover in the individuals occupying particular rocks and riffles because of continuing movement and dislodgement. Changes in species composition were frequently observed for individual rocks and whole riffles and occasionally for entire cave streams (Culver 1973a). The chance events of migration and extinction should result in some habitable patches being unoccupied, and it was observed that most rocks, some riffles, and occasionally whole streams were unoccupied.

The model used (Culver 1973a) allows the proportion of rocks occupied to be a balance between births and emigrations from other patches, on the one hand, and washouts, on the other. If p_i is the frequency of rocks occupied by species i, then

$$\frac{dp_i}{dt} = m_i p_i (1 - p_i) - e_{ii} p_i^2 - \sum_{j \neq i} e_{ij} p_i p_j \qquad (6\text{-}8)$$

The first term is the birth plus emigration rate, m_i, times the frequency of occupied spaces (p_i) times the frequency of unoccupied spaces ($1 - p_i$). The second term is the washout rate due to an intraspecific collision (e_{ii}) times the frequency of intraspecific collisions (p_i^2), and the remaining terms are the analogous effects of interspecific collisions on washout rate. Equations of this type have been criticized by Levin (1974) and Slatkin (1974) because the probability of contact between two species is in general not equal to the product of their separate frequencies, in other words, $p_i p_j$. But for this particular system the short time scales and high frequency of mixing greatly reduce this problem. Equation 6-8 can be rearranged in the form of standard competition equations as follows:

$$\frac{dp_i}{dt} = m_i p_i \frac{K_i - p_i - \sum_{j \neq i} \alpha_{ij} p_j}{K_i} \tag{6-9}$$

where

$$K_i = \frac{m_i}{m_i + e_{ii}} \quad \text{and} \quad \alpha_{ij} = \frac{e_{ij}}{e_{ii} + m_i} \tag{6-10}$$

The washout rates, e_{ii} and e_{ij}, can be measured directly in the laboratory, using appropriate combinations of species and controlling for total density. The birth and emigration rates cannot be measured directly, but m_i must be small compared to e_{ii} because the frequency of rocks occupied by a species when alone ($p_i = m_i/m_i + e_{ii}$) is small. With the arbitrary assumption that $m_i = 0.01$, competition coefficients, α_{ij}, were determined for three Powell Valley species (Culver 1973a):

$$\begin{bmatrix} 1 & \alpha_{12} & \alpha_{13} \\ \alpha_{21} & 1 & \alpha_{23} \\ \alpha_{31} & \alpha_{32} & 1 \end{bmatrix} = \begin{bmatrix} 1 & 0.99 & 1.32 \\ 0.32 & 1 & 1.29 \\ 1.16 & 0.49 & 1 \end{bmatrix} \tag{6-11}$$

with *Crangonyx antennatus* as species 1, *Caecidotea recurvata* as species 2, and *Lirceus usdagalun* as species 3. Similarly, competition coefficients were found for two Greenbrier Valley species:

$$\begin{bmatrix} 1 & \alpha_{12} \\ \alpha_{21} & 1 \end{bmatrix} = \begin{bmatrix} 1 & 2.46 \\ 5.68 & 1 \end{bmatrix} \tag{6-12}$$

with *Caecidotea holsingeri* as species 1 and *Caecidotea scrupulosa* as species 2.

Predicted Microhabitat Separation The competition coefficients calculated by formula 6-10 are not in any sense niche overlaps. Rather, they purport to be measures of the intensity of competition and should be positively correlated with the amount of microhabitat separation. By far the largest α values are associated with *Caecidotea holsingeri* and *C. scrupulosa*. The latter is known from twelve caves in the southern part of the Greenbrier Valley (Monroe and Greenbrier counties), and *C. holsingeri* is known from sixteen caves in the same area. The two have been found in the same cave (General Davis Cave) only once, and even then they were not found at the same time or place in the cave. For the Powell Valley species, there is a match between intensity of competition and amount of microhabitat separation. The species pair that competes the least is *Caecidotea recurvata* and *Crangonyx antennatus,* and they are routinely found in the same riffle, with the smaller *C. antennatus* deeper in the gravels. The species pair that competes the most is *Lirceus usdagalun* and *C. antennatus,* and they never coexist in the same stream. The maximum amount of microhabitat overlap observed for these two species occurs where *C. antennatus* is limited to side pools off the main stream (Fig. 6–9). The pair with intermediate competition is *C. recurvata* and *L. usdagalun,* and they occur in different riffles of the same stream.

Stability Rules It is also possible to analyze the stability of the equilibrium of various species combinations using the competition coefficients. Since all three species co-occur in three caves (See Table 6–3), the calculated values of α_{ij} should result in a stable equilibrium, which they do. Following Lawlor (1980), one can ask whether the observed minimum eigenvalue of the observed matrix of competition coefficients is significantly different from minimum eigenvalues obtained from matrices with the same elements as the observed matrix but arranged at random. For the α-matrix, the minimum eigenvalue must be positive for stability. The mean minimum eigenvalue of 100 such randomly constructed matrices is -0.12, with a standard deviation of 0.20, indicating that most randomly arranged communities are unstable. The

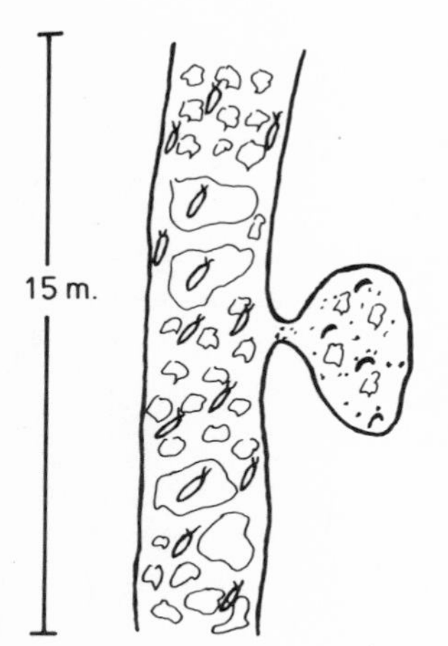

Figure 6–9 Schematized map of the distribution of *Lirceus usdagalun* (oblong symbols) and *Crangonyx antennatus* (crescent symbols) in Surgener Cave, Lee County, Virginia. The large irregular shapes represent large rocks (>10 cm) and the small irregular shapes represent small rocks in a riffle. The side pool is mud-bottomed with a few small rocks. In subsequent visits, *C. antennatus* had disappeared, and *L. usdagalun* was in the side pool. (From Culver 1973a; copyright 1973, the Ecological Society of America.)

minimum eigenvalue of the actual matrix is $+0.08$. The minimum value is larger than the average of the randomly constructed matrices but not significantly larger.

More generally, it is possible to use the calculated α values and estimates of the carrying capacities, K_i, to determine which combinations of species are stable and which species can invade which communities (Culver 1976). Two species pairs are predicted to be unstable when the pair is isolated from some third species: *L. usdagalun* with *Crangonyx antennatus*, and *L. usdagalun* with *Caecidotea recurvata*. Neither of these pairs has been found in isolation in any cave stream, the closest case being the one shown in Figure 6–9, which did not persist.

The absence of predicted unstable pairs can be made somewhat more quantitative. If there are such assembly rules, as Diamond (1975) termed them, then unstable communities should occur less frequently, on a statistical basis, than expected by chance. The actual analysis is complicated by the small range of *L. usdagalun*. Inclusion of cave streams that this species has never reached would obscure the analysis, so the following rather arbitrary convention was adopted: a cave stream was included if it was within one km of a known locality of *L. usdagalun*. Table 6–4 summarizes the results of the analysis. Because of the small number of caves, the results are only marginally significant, even though no caves had "forbidden" communities.

These results point up a recurring problem in cave ecology. The rela-

Table 6–4 Observed communities and subcommunities of *Caecidotea recurvata* (Cr), *Crangonyx antennatus* (Ca), and *Lirceus usdagalun* (Lu) in cavestreams within the geographic range of all three. Stable and unstable combinations were determined by stability analysis of α and K values determined in a laboratory stream. Expected numbers were generated by assuming that species were distributed at random.

Stable combinations			Unstable combinations		
Species	Observed	Expected	Species	Observed	Expected
None	0	0.16	Lu-Cr	0	1.02
Lu	2	0.41	Lu-Ca	0	1.02
Cr	0	0.41			
Ca	0	0.41			
Cr-Ca	2	1.02			
Cr-Ca-Lu	3	2.55			

$\chi_1^2 = 2.80$, $p > 0.90$

tive simplicity of the systems makes possible more complete predictions than is usually the case. For example, in this situation there is a complete *a priori* set of assembly rules for the aquatic community. But the very simplicity of the community, and in this case the restricted ranges of the species, makes statistical testing difficult. There is much to be said on both sides of the argument (see Culver 1978 and Pimm 1978). My own point of view is that we lose much by taking the overly dogmatic view that the only interesting results are statistically significant, but the absence of firm statistical results makes any conclusions more in the way of suggestions than conclusions.

Qualitative data on distributions and on successful and failed invasions lend weight to the existence of assembly rules. Especially impressive is the species distribution pattern in the three physically distinct sections of the cave stream in Thompson-Cedar Cave. All three species occur in the downstream section, *C. recurvata* and *C. antennatus* occur in the upstream section, and *L. usdagalun* occurs alone in the middle section. Thus no predicted unstable communities occur even with the species in very close proximity. Also as predicted, neither *C. antennatus* nor *C. recurvata* has been able to successfully invade a stream dominated by *L. usdagalun*. Only one successful inva-

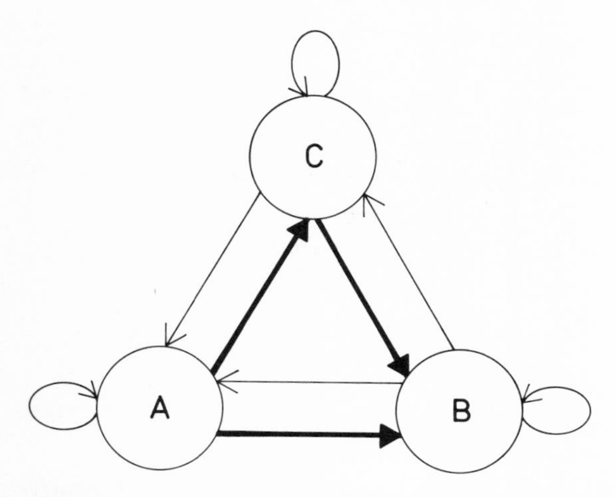

Figure 6–10 Illustration of the direct effect and the indirect effect of species A on species B. Each arrow indicates a competitive effect. In this example all three species compete with each other. The arrows from a species back to itself represent intraspecific competition. The heavy arrow from A to B is the direct negative effect. The heavy arrows from A to C and back to B represent the indirect positive effect of species A on species B.

sion has been recorded; as predicted, *L. usdagalun* successfully invaded a community with the other two species (Culver 1976).

Apparent Mutualisms and Indirect Effects Correlations and partial correlations of species through time provide further insights into the community. Levins (1975) and Levine (1976) have pointed out that when three or more competitors are present, there is the possibility that some pairs of competitors will be positively correlated and be "apparent mutualists," rather than negatively correlated as is expected intuitively. This can arise in the following way, as shown in Figure 6–10. When there are three competitors, species A has two effects on species B: first, a direct negative effect, and second, an indirect positive effect via species C. That is, species A has a negative effect on species C, which has a negative effect on species B, so the overall indirect effect of species A on species B is positive. The correlation between the two species (A and B) depends on the relative magnitude of these effects. Davidson (1980) has documented a very striking case of an apparent mutualism between competitors in a desert ant community.

The actual computations of indirect effects are rather lengthy (Levine 1976), so I will give only the bare essentials here. A change in the growth equations, $f_i = dN_i/dt$, can be written as $\partial f_i/\partial C_h$ where C_h is some parameter that affects the growth equation. This partial derivative can be thought of as a change in growth rate of a species due to a change in the species carrying capacity. These changes will in turn affect population size of species i $(\partial N_i/\partial C_h)$ in the following way if the population is near equilibrium (Levins 1975). For three competitors:

$$\frac{1}{|A|}\begin{bmatrix} (a_{22}a_{33} - a_{23}a_{32}) & (-a_{33}a_{12} + a_{32}a_{13}) & (-a_{22}a_{13} + a_{12}a_{23}) \\ (-a_{21}a_{33} + a_{23}a_{31}) & (a_{11}a_{33} - a_{13}a_{31}) & (-a_{11}a_{23} + a_{13}a_{21}) \\ (-a_{22}a_{31} + a_{21}a_{32}) & (-a_{11}a_{32} + a_{12}a_{31}) & (a_{11}a_{22} - a_{12}a_{21}) \end{bmatrix} \begin{bmatrix} \dfrac{-\partial f_1}{\partial C_h} \\ \dfrac{-\partial f_2}{\partial C_h} \\ \dfrac{-\partial f_3}{\partial C_h} \end{bmatrix} = \begin{bmatrix} \dfrac{\partial N_1}{\partial C_h} \\ \dfrac{\partial N_2}{\partial N_2} \\ \dfrac{\partial N_3}{\partial C_h} \end{bmatrix} \qquad (6\text{-}13)$$

where a_{ij} is the effect of species j on species i, and $|A|$ is the determinant of the matrix of interaction coefficients, a_{ij}. For the standard competition equations,

$$a_{ij} = \frac{-r_i \hat{N}_i}{K_i} \alpha_{ij} \tag{6-14}$$

Each term of the matrix is the effect of a change in carrying capacity of species i on the population size of species j. Each off-diagonal term (a'_{ij}) is the difference between the direct and the indirect effect for species j on species i. Each diagonal term (a'_{ii}) is the determinant of the subcommunity formed by deleting species i.

Correlations can be predicted in the following way. A change in K_i (or any parameter affecting species i directly) results in changes in population sizes of all species; these are given in column i of equation 6-13. The expected correlation between two species because of changes in K_i can be found by comparing the signs of the two appropriate terms in column i. Using the above recipe, the elements of the matrix for the Lee County cave communities have the following signs:

Effect on population size of	Increase in K of		
	C. antennatus	*C. recurvata*	*L. usdagalun*
C. antennatus	+	−	0
C. recurvata	+	−	−
L. usdagalun	−	+	+

A positive correlation is expected between species i and species j when the product of the elements $a'_{ik} \times a'_{jk}$ of the matrix in equation 6-13 is zero or has the same sign for each value of k, in other words, for the entire row. Thus, *C. antennatus* and *C. recurvata* should be positively correlated.

In contrast with the correlation coefficients, all partial correlations of competitors should be negative because of the close correspondence between the definitions of partial correlation and of competition coefficients. Both measure the effect of species j on species i with all other species held constant.

The set of predictions and the actual data (see Culver 1981) are given in Table 6–5. Of the nineteen predictions about the signs of cor-

Table 6–5 Comparison of observed and predicted correlations and partial correlations of species abundance through time. (Culver 1981.)

		Partial correlation			Correlation		
Cave	Species pair	Predicted	Observed	*p*	Predicted	Observed	*p*
Spangler	*C. recurvata* *C. antennatus*	–	−0.34[1]	N.S.	–	−0.34[1]	N.S.
Gallohan No. 2	*C. recurvata* *C. antennatus*	–	−0.99	>.99	+	+0.28	N.S.
	C. antennatus *L. usdagalun*	–	−0.99	>.99	–	−0.57	N.S.
	C. recurvata *L. usdagalun*	–	−0.99	>.99	–	−0.95	>.95
Gallohan No. 1	*C. recurvata* *C. antennatus*	–	+0.30	N.S.	+	+0.11	N.S.
	C. antennatus *L. usdagalun*	–	−0.82	N.S.	–	−0.82	>.95
	C. recurvata *L. usdagalun*	–	+0.05	N.S.	–	+0.07	N.S.
Court Street	*C. holsingeri* *S. emarginatus*	–	−0.36	N.S.	–	−0.38	N.S.
	C. holsingeri *S. spinatus*	–	−0.51	N.S.	–	−0.48	N.S.
	S. emarginatus *S. spinatus*	–	−0.01	N.S.	+	+0.19	N.S.

1. Correlation and partial correlation are identical because there are only two species in the community.

relations and partial correlations, sixteen are in agreement with the signs of the calculated values. This level of agreement would be attained on a chance basis with a probability of only 0.002 (sign test). In addition, five of the correlations were statistically significant. Especially interesting is the complete agreement of observed and predicted correlations and partial correlations for Gallohan Cave No. 2. The time period of sampling in Gallohan No. 2 covers the period of the invasion of *L. usdagalun* (Dickson 1976), a time of intense competition but also a time when the populations are far from equilibrium. This suggests that the linear models used may hold when the situation is far from equilibrium.

The interaction between *Crangonyx antennatus* and *Caecidotea recurvata* epitomizes the importance of indirect effects in the organization of communities. When no other competitors are present (Spangler Cave) the two species are negatively correlated. When a third competitor is present (Gallohan Cave No. 1 and No. 2), the two species are positively correlated.

Effects of Predation In a few caves in the Powell Valley, larvae of the salamander *Gyrinophilus porphyriticus* are important predators of the amphipods and isopods. From the point of view of the behavioral ecologist, their feeding behavior is very simple (Culver 1973b) and even uninteresting. The larvae live in pools in the stream, and when hungry they rise up on their front legs and also usually their hind legs. When an amphipod or isopod comes within 2 to 4 cm of its snout, the larva will eat the prey with a rapid sucking action. The larvae apparently do not distinguish between prey species, but they are more successful in capturing isopods than amphipods, because amphipods sometimes escape by swimming off, while isopods do not. At least in the laboratory, the probability of successful predation by the larvae is quite high: over 75 of their feeding attempts on *Caecidotea recurvta* were successful. Peck (1973b) found similarly high successful predation rates in caves for the salamander *Haideotriton wallacei*. Individuals in riffles suffer little predation because almost all the larvae are in quieter waters, where water currents interfere little with their mechanoreception of prey movements. In McClure's Cave, the most intensively studied cave, actual predation rates on the two prey species, *C. recurvata* and *C. antennatus*, are very similar to the laboratory results. Although the probability of successful capture of *C. antennatus* is lower, the fraction of the population accessible to predation is higher, because it is partly excluded from riffles by *C. recurvata* (Culver 1975). The functional response (number of prey taken plotted against prey density) is linear over the naturally occurring range of prey densities. The absence of the usual nonlinearities in the functional response curve results from the negligible handling times of prey and the absence of any evidence of predator satiation.

In spite of nearly equal predation rates and a linear functional response curve, predation models have been of very limited use in answering two questions: first, how does predation affect the stability and size of prey populations? and second, why don't most cave invasions by cave salamanders result in successful establishment of a salamander population? The reasons for the limited utility of models is very different for the two questions. Analysis of a model of the two competitors (*C. antennatus* and *C. recurvata*) with equal predation rates on the two indicates that predation either stabilizes or destabilizes the system, depending on whether the intrinsic rate of increase of *C. antennatus* is greater or less than the intrinsic rate of increase of *C. recurvata* (Culver 1975). Neither of these parameters has been measured, and there is little likelihood they will be, given the very low rates of increase of most cave populations (chapter 3). In the case of field observations indicating that invasion rarely results in establishment, the predation model is actually misleading in a way that will be discussed below.

In McClure's Cave, predation results in an increase in the density of *C. antennatus* and a decrease in the density of *C. recurvata* in the immediate vicinity of salamander larvae (Fig. 6–11A). Since population densities of *C. antennatus* are low when the predator is absent, predation stabilizes the system in the sense that *C. antennatus* is less likely to become extinct because of random fluctuations in population size. In sharp contrast, in Sweet Potato Cave *C. recurvata* is absent in the vicinity of predators and *C. antennatus* is reduced in density (Fig. 6–11B). The differential effects of predation in the two caves results from the differences in the physical environment. In Sweet Potato

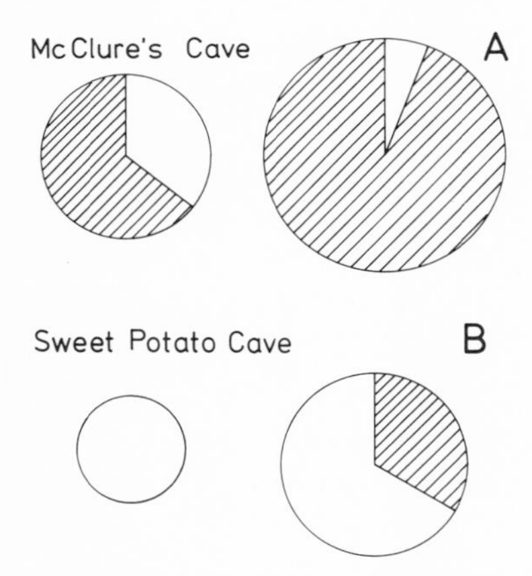

Figure 6–11 Effect of predation by larval *Gyrinophilus porphyriticus* on its prey in (*A*) McClure's Cave and (*B*) Sweet Potato Cave. The area of the circle indicates total prey abundance. The striped sector represents *Caecidotea recurvata;* the open sector represents *Crangonyx antennatus*. The circles on the left represent prey abundance in the presence of a predator; those on the right represent prey abundance in the absence of a predator.

Table 6–6 Fraction and density (per 0.09 m^2) of the prey populations of McClure's Cave accessible to salamander predators when predators are nearby and when they are not. (From Culver 1975.)

Prey	*Gyrinophilus* larvae nearby	Fraction accessible to predator	Density accessible to predator
Crangonyx antennatus	Yes	0.29	0.30
	No	0.51	0.15
Caecidotea recurvata	Yes	0.10	0.13
	No	0.19	0.78

Cave the habitat is a series of mud-bottomed rimstone pools. There are no refugia for the prey in the form of riffles. Predation is high enough so that isopods are not present in the same pool with salamander larvae. In some pools *C. antennatus* persists with salamanders because amphipods are harder to capture and because they burrow in the mud (Holsinger and Dickson 1977). In contrast, in McClure's Cave only a fraction of each prey population is vulnerable to predation, because the salamanders do not occur in riffles or flowstone habitats.

In McClure's Cave a comparison of areas near and away from larvae showed that both the fraction and the density of the *C. recurvata* and the fraction of *C. antennatus* accessible to predators declined in the vicinity of a predator (Table 6–6). Consequently, colonizing salamanders encounter a relatively dense, accessible prey population, and initially the predation rate for each larva is higher than for an individual in a resident population, where the prey are relatively scarce and inaccessible.

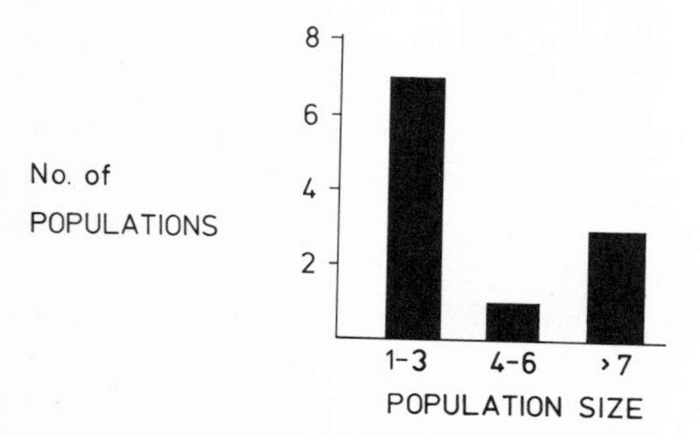

Figure 6–12 Frequency distribution of numbers of larval *Gyrinophilus porphyriticus* in caves in Lee County, Virginia, and Claiborne County, Tennessee. (From Culver 1975.)

Therefore, predation is not constant, as is assumed in the model. Biologically, we would expect most "populations" of predators to consist of a few individuals, since population growth would be difficult, and this is in fact the case (Fig. 6–12). Only three of eleven caves have populations of sufficient size to expect a persistent population.

Evolutionary Considerations There is some evidence that the intensity of competition between pairs of species declines with evolutionary time. The length of time each Appalachian isopod and amphipod has been in caves can be roughly determined by examining their distribution patterns and the amount of regressive evolution, which should increase with time (Culver 1976). The "cave age" of the species can be conveniently divided into four groups. The youngest species occur in cold-water habitats outside caves and show little signs of regressive evolution, retaining eyes and pigment. In the next group are species with very restricted ranges and with reduced eyes and pigmentation. Species in the third group have large geographic ranges, no pigment, and vestigial eyes. In the oldest group are species without eyes and pigment, with small ranges per species but with large species groups in which the species show clear morphological differences. Further justification of this scheme is given in Culver (1976).

With one exception, the length of time available for interaction between two species can be estimated by the cave age of the younger of the pair. The exception is a pair of old species (*Caecidotea recurvata* and *C. richardsonae*) whose ranges are barely overlapping and whose contact is much more recent than their cave age. Intensity of interaction is known directly for those species pairs whose α's were measured, and indirectly for a larger set of species on the basis of microhabitat separation, which is correlated with the intensity of competition (Culver 1973a). The results are shown in Table 6–7, with *C. recurvata–C. richardsonae* deleted for the reasons given above. There is a perfect and statistically significant rank-order correlation between average minimum age of interaction and amount of separation. The rank-order correlation of $\alpha_{ij} \cdot \alpha_{ji}$ with age of interaction is suggestive but only marginally significant (Culver 1978).

The reduction in intensity of competition between two species over time does not mean that any species experiences less overall competition with time. With invasions of new species into caves and migrations of other cave species, there is no evidence that the overall amount of competition experienced by a species diminishes.

To many field biologists, character displacement is the *sine qua non* of interspecific competition. Yet it has been repeatedly shown by theo-

Table 6–7 Intensity of competition and relative length of isolation in caves. (From Culver 1976, reprinted by permission of the University of Chicago Press, © 1976, the University of Chicago.)

Species pair	Valley	Relative age of interaction[a]
Species in different habitats of same riffle		
Gammarus minus-Stygobromus spinatus	Greenbrier	1
Stygobromus emarginatus-Stygobromus spinatus	Greenbrier	4
Stygobromus emarginatus-Caecidotea holsingeri	Greenbrier	4
Stygobromus spinatus-Caecidotea holsingeri	Greenbrier	4
Gammarus minus-Caecidotea holsingeri	Greenbrier	1
Crangonyx antennatus-Caecidotea recurvata[b] ($\alpha_{ij}\alpha_{ji} = 0.3$)	Powell	3
$\bar{X} = 2.8$		
Species in different riffles		
Crangonyx antennatus-Caecidotea recurvata[b]	Powell	3
Lirceus usdagalun-Caecidotea recurvata ($\alpha_{ij}\alpha_{ji} = 0.65$)	Powell	2
Caecidotea scrupulosa-Gammarus minus[c]	Greenbrier	1
Caecidotea scrupulosa-Crangonyx sp.	Greenbrier	1
$\bar{X} = 1.8$		
Species barely coexisting		
Caecidotea scrupulosa-Gammarus minus†	Greenbrier	1
Gammarus minus-Stygobromus emarginatus	Greenbrier	1
Crangonyx antennatus-Lirceus usdagalun ($\alpha_{ij}\alpha_{ji} = 1.5$)	Powell	2
$\bar{X} = 1.3$		
Complete exclusion		
Caecidotea scrupulosa-Caecidotea holsingeri ($\alpha_{ij}\alpha_{ji} = 13.4$)	Greenbrier	1
Crangonyx sp.-*Gammarus minus*	Greenbrier	1
$\bar{X} = 1.0$		

a. See text for explanation of relative age. The larger the number, the longer the time species has been isolated in caves.

b. *C. antennatus* and *C. recurvata* are usually found in same riffle, but in a few caves (*e.g.*, Cope Cave) they are in different riffles.

c. Either *C. scrupulosa* and *G. minus* are in different riffles, or *C. scrupulosa* is very rare.

retical ecologists (MacArthur and Levins 1967; Slatkin 1980) that competition can result in character convergence as well as character displacement. Generally, competition can be reduced by habitat selection or by difference in foraging times. Evidence for or against character displacement is not evidence for or against competition, but evidence for or against the particular kind of competition that leads to character displacement.

There are no documented cases of character displacement among aquatic cave invertebrates, but one apparent case of character displacement is worth considering. Two closely related species of isopods, *Caecidotea cannula* and *C. holsingeri*, occur in caves in

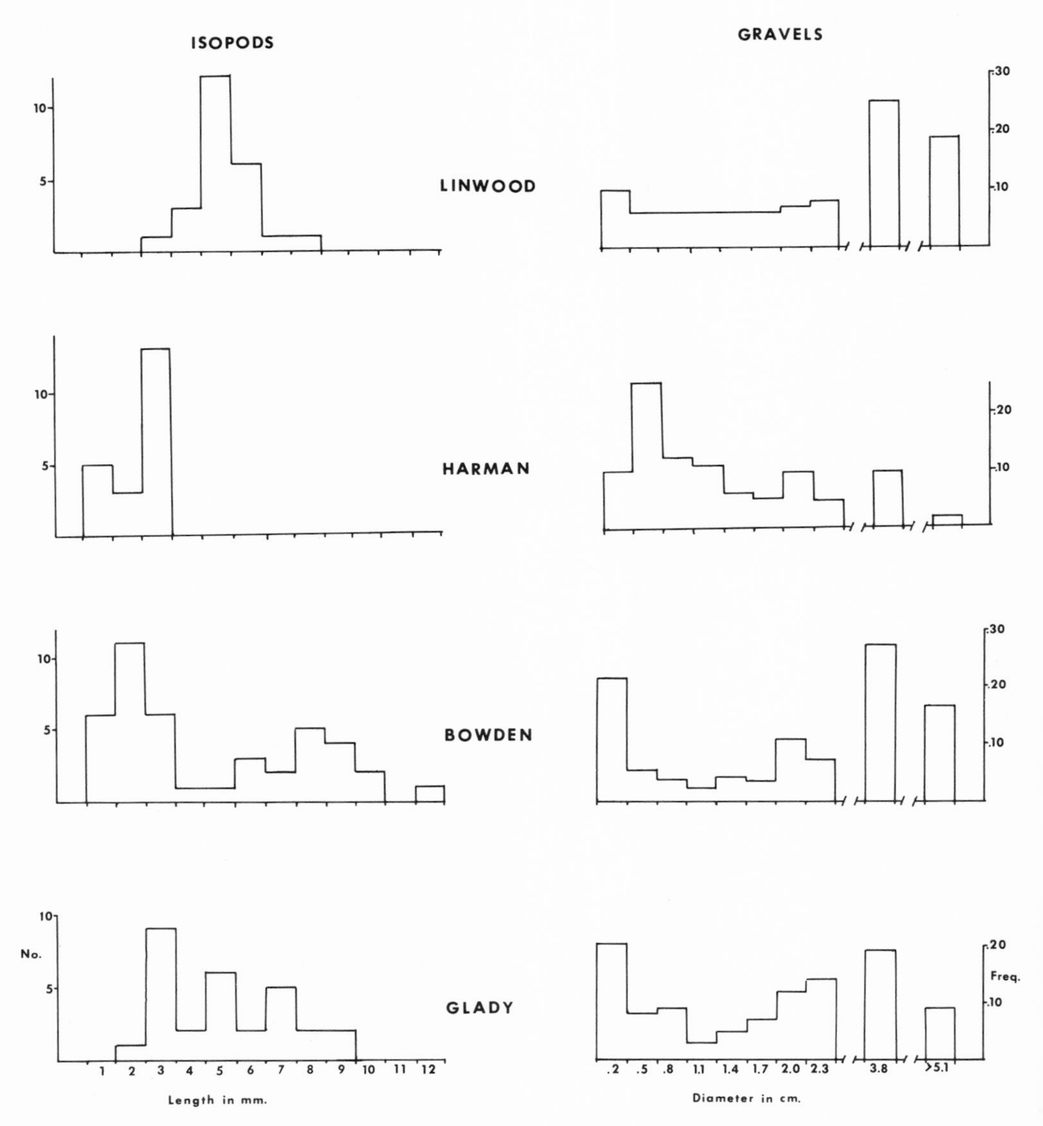

Figure 6–13 Gravel and isopod sizes for Linwood, Harman, Bowden, and Glady Caves, West Virginia. See Table 6–8 for a description of the distribution. (From Culver and Ehlinger 1980.)

Table 6–8 Qualitative characteristics of the isopod and gravel size distributions shown in Figure 6–13. Quartiles and medians are given for each distribution (length in mm for isopods, diameter in mm for gravels). Frequency (F) of large gravel (>12.3 mm) is also given. (From Culver and Ehlinger 1980.)

	Gravel					Isopods			
Cave	Q_1	M	Q_3	Qualitative features	F	Q_1	M	Q_3	Qualitative features
Linwood	5.6	11.9	19.1	Uniform for small gravel	0.44	5.0	5.4	6.2	Strongly unimodal, narrow size range, all intermediate in size
Harman	2.4	5.6	10.3	Unimodal, skewed to small sizes	0.12	1.6	3.1	3.6	Unimodal, narrow size range, all small in size
Bowden	2.4	11.9	19.1	Strongly bimodal	0.43	2.4	3.3	8.3	Bimodal, broad size range
Glady	2.4	8.7	19.1	Weakly bimodal	0.27	3.7	5.3	7.6	Unimodal, broad size range, skewed toward small size

northern West Virginia. In the Monongahela River drainage where both occur, *C. cannula* is twice the size of *C. holsingeri,* and in other drainage systems where *C. cannula* does not occur, *C. holsingeri* is larger, which led us to suspect character displacement (Holsinger, Baroody, and Culver 1975). We also suspected that the size of gravels in a cave stream had a strong effect on isopod size. Experiments in a laboratory stream indicated that both hypotheses were possible: isopods of different sizes competed less strongly than isopods of the same size, and large isopods could better maintain their position in a current with large gravels than with small gravels. The question is which factor is more important in the field. There is a strong correlation between the shape of the distribution of gravel sizes and the shape of the distribution of isopod sizes (Fig. 6–13, Table 6–8), indicating that isopod size is largely determined by gravel size in the stream rather than by the presence of competitors (Culver and Ehlinger 1980). The two mismatches of distributions provide no support for character displacement. In Glady Cave, gravel sizes are bimodally distributed, while isopod sizes are unimodal, but if character displacement were important, the differences in isopod size should be more rather than less pronounced. The mismatch in Linwood Cave results from the absence of *C. cannula* from the Elk River drainage rather than from competitive effects.

Summary: The Role of Models

The differential equation models used in the previous sections allowed a deeper probing into the structure and dynamics of cave stream communities than otherwise would have been possible. It was possible to proceed in a logical way toward an explanation of the varying amounts of microhabitat separation between species and to successfully predict stable and unstable species combinations. The models suggested some nonobvious patterns to search for, especially the relative constancy of total abundance and the possibility of indirect mutualisms. It was also clear when the models were inadequate. The predation model used was insufficient to explain some important features of the predator–prey system, most notably niche shifts of the prey and the relative ease of predator invasion compared to predator establishment.

Two factors stand in the way of making a very strong claim about the importance for noncave systems of the particular models used for caves. The first is the problem of statistical testing. Because of the very simplicity of a community that allows measuring pairwise interactions in the first place, the data base is relatively small. The most extreme case occurred in the attempt to test assembly rules. No unstable subcommunities were found, yet the results were only marginally significant. In some cases it was possible to generate a larger data base, but it is unlikely that cave ecologists in general will be able to generate the large data sets that it is currently in vogue to test. What data from cave communities can provide is a clear indication of whether a particular model seems to work in a relatively simple system.

The second objection raised about cave communities is that they are in some way so aberrant that the rules governing their structure and dynamics are either completely different from those for other communities or so trivial as to not constitute worthy objects of study. What is obviously missing from caves are green plants and thus plant–animal interactions. However, the study of other detritus-based communities, such as freshwater streams, has added greatly to our general ecological knowledge. Other interactions clearly are present in caves, including competition, predation, and symbiosis. Free-living mutualists may or may not be present, but that statement can be made for many temperate zone communities.

Even the relatively simple aquatic stream communities of Appalachian caves make clear some inadequacies in the current questions being asked about species interactions. It is difficult to state precisely what it means to weigh the relative importance of different species interactions. Consider the communities with predaceous *Gyrinophilus porphyriticus* larvae and its prey, *Caecidotea recurvata* and *Crangonyx antennatus*. The predator clearly alters the relative abundance of prey and can cause the extinction of *C. recurvata* in particular habitats, but the predator effect just as clearly depends on competition between the prey. Both interactions are important, and the interactions are complex enough that these complexities, which Levins (1975) termed network effects, have come to predominate in the community. Network effects are important not only for the predator–prey systems but also for three competitors, where indirect mutualisms were observed.

The potential of cave communities for studies of species interactions is by no means exhausted. In particular, terrestrial communities are generally more complicated than aquatic communities, and their potential is largely untapped. The terrestrial fauna of Mammoth Cave has attracted attention because of its complexity (see Barr 1967a), but even the supposedly simple terrestrial faunas of most caves is more complex than those of aquatic communities. Shelta Cave in Alabama has one of

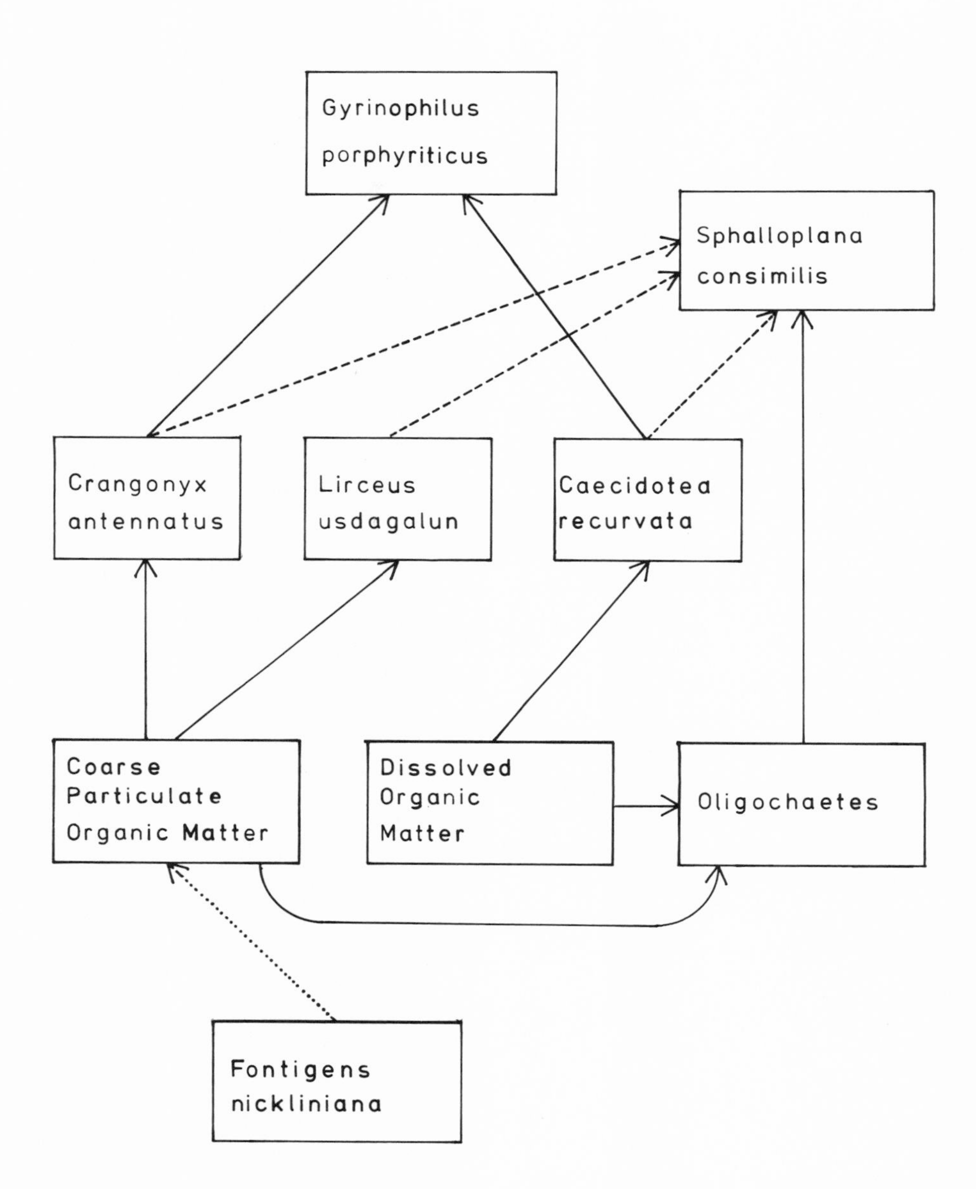

Figure 6–14 Food web for the aquatic fauna of Gallohan Cave No. 1, Lee County, Virginia. Dashed lines indicate feeding on dead and moribund individuals; dotted lines are conjectured feeding relationships.

Table 6–9 Visual census of terrestrial arthropods in a 100 m by 5 m strip of wet passage with organic debris in Gallohan Cave No. 1. Species marked with an asterisk are cave-limited (troglobites). (Data from T. Kane, unpublished.)

Class	Order	Species	No. individuals
Crustacea	Isopoda	*Amerigoniscus henroti**	2
Arachnida	Acarina	*Rhagidia* sp.*	1
	Araneae	*Nesticus carteri*	11
		*Phanetta subterranea**	3
	Pseudoscorpionida	*Kleptochthonius proximosetus**	2
Diplopoda	Chordeumida	*Pseudotremia nodosa**	47
		Pseudotremia valga	5
Insecta	Diplura	*Litocampa cookei**	1
	Collembola	*Pseudosinella orba**	12
		Tomocerus bidentatus	2
		Arrhopalites hirtus	4
	Coleoptera	*Pseudanophthalmus delicatus**	13

the richest aquatic faunas in North America, yet there are as many terrestrial troglobitic species, twelve, as aquatic troglobitic species (Cooper 1975). Gallohan Cave No. 1 in Lee County, Virginia, has an exceptionally rich aquatic fauna, the food web of which is shown in Figure 6–14. By contrast, the less thoroughly studied terrestrial fauna includes at least twelve species (Table 6–9), the feeding relationships of which are nearly a complete mystery.

7 Zoogeography

There have been two very different approaches to the study of the distribution of cave organisms. The historical approach concentrates on the factors responsible for the invasion and isolation of organisms in caves, such as the warming and drying associated with Pleistocene interglacials, and the subsequent movements of cave organisms. The island biogeographic approach concentrates on the potential analogy between caves and islands, emphasizing the dynamics of migration and extinction.

These two approaches are competing hypotheses, but they are also hypotheses to explain different phenomena. Historical biogeographers deal with relatively large regions such as the Appalachians over geologic time, whereas island biogeographers deal with small areas such as the caves in one drainage basin over a short time period. But the hypotheses clearly infringe on each other. In a historical context, the extinctions studied by the island biogeographer destroy history, and the migrations obscure it. In an island context, historical patterns may obscure dynamics. Finally, they are competing hypotheses when both are used on the same geographic scale.

This chapter has three main sections. The first considers the large-scale patterns deduced by historical biogeographers, in particular, the causes of isolation in caves and evidence concerning subsequent underground dispersal. The second section considers the validity of the caves-as-islands analogy and at what geographical scale, if any, it is valid. The third section considers the predictions generated by the two approaches for the same data set: species distributions in caves in the Kanawha and Elk River drainages in West Virginia.

For many topics discussed in the preceding chapters there was a real paucity of relevant data. That is not the case here. Although many cave-limited species have not yet been described, the cave species of many invertebrate groups are better known than the surface species, for instance, spiders (Deeleman-Reinhold 1981), millipeds (Shear 1972), and nearly all of the Mexican cave fauna (Reddell 1981). There are species lists for large numbers of caves in Europe and North America. For example, Reddell (1981) lists the troglobitic fauna (more than 250 species) of over 500 Mexican caves, and Holsinger, Baroody, and Culver (1975) list the invertebrate fauna (188 species) of 190 caves in West Virginia. Although there is no cave region in which all the caves have been biologically explored, as much or more data are available on detailed species distribution in caves than for almost any other habitat. Recently, $^{230}Th/^{234}U$ ratios have been used to determine the age of speleothems in caves and $^{18}O/^{16}O$ ratios of speleothems to measure paleotemperatures (Thompson, Schwartz, and Ford 1974). These techniques allow for relatively precise determination of cave ages and determination of the length and number of interglacial periods, which are probable times of invasion and isolation of terrestrial organisms (see below).

In spite of these advantages, many cave biogeographers have been reluctant to go beyond the description of new species and the delineation of ranges. There are several reasons for this. The absence of a useful fossil record for most of the ancestors of cavernicoles has made it difficult to speculate about historical biogeography. Ostracods are an exception, and Danielopol (1980) has made good use of the ostracoid fossil record in explaining the origin and distribution of the European subterranean ostracods. A more important reason for the lack of speculation about the time of isolation and the subsequent movements of cave organisms is that until recently this was not the primary question being asked. A group of French cave biologists, especially Jeannel and Vandel, adhered to an orthogenetic theory of evolution, which, among other things, contended that organisms in caves were in senescent, and therefore very old, phyletic lines. They focused attention not on the time of invasion and isolation in caves, but rather on the time of origin of the supraspecific category being studied. They attempted to show, either through the fossil record or by the absence of any relatives in surface habitats, that cave organisms are in phyletically old taxa. While this is undoubtedly the case for some cave organisms, such as the amphipod family Crangonyctidae, European investigators have shown that it is not generally true (see Deeleman-Reinhold 1981 and Danielpol 1980). Orthogenesis attracted few if any English-speaking biologists, but it apparently had a stultifying effect on biogeographic speculation.

Invasion and Isolation in Caves

Terrestrial Species For terrestrial cave organisms in temperate-zone caves, the general consensus (Barr 1968, Peck 1981a) is that their immediate ancestors are forest soil- and litter-dwelling arthropods that invaded caves and were isolated there during Pleistocene interglacials. There is no direct fossil record evidence for this, but there is strong indirect evidence. First, caves in glaciated areas have a very depauperate or nonexistent terrestrial fauna. Second, the closest surface relatives of many cave-limited species are found in the litter of boreal forests. Perhaps the most striking example is the trechine beetle *Pseudanophthalmus sylvaticus*, the only North American surface-dwelling species in a genus with over 200 North American species. *P. sylvaticus* is known only from the underside of rocks embedded in stream gravels in a mixed hemlock and deciduous forest at an elevation of 1,000 meters in West Virginia (Barr 1967b). Third, the regions most affected by the Pleistocene, short of actually being ice-covered, have the largest number of terrestrial troglobites. In Mexico most terrestrial troglobites are from high-altitude caves, where the effects of the Pleistocene were presumably most pronounced (Reddell 1981). Fourth, regions lacking forests throughout the Pleistocene, such as parts of the American Southwest and Australia (Peck 1980), have a very depauperate terrestrial cave fauna. Fifth, estimates of time since the divergence of closely related species, using Nei's (1975) formula and based on electrophoretic data, are consistent with isolation in the Pleistocene (Delay et al. 1980).

This sort of analysis can be carried further by attempting to determine which species were isolated in which interglacial. The intermediate level of regressive evolution and morphological modification of Collembola in the Driftless Area of Iowa, Illinois, Wisconsin, and Minnesota, which was not covered by the Wisconsin glaciation, indicates that the more highly modified Collembola farther south were isolated in earlier interglacials. Peck has made an admirable attempt to pinpoint which interglacials were responsible for isolation of particular arthropod species in Grand Canyon caves (Peck 1980) and among the species in the *Ptomaphagus hirtus* group (Peck 1981b). Both studies are worth examining in more detail.

The Grand Canyon caves he studied are at elevations between 1,160 m and 1,580 m. Present forests are on the Kaibab Plateau between 2,000 m and 2,800 m. Peck reports that according to radiocarbon dating of packrat middens, the vegetation zones were lowered as much as 1,000 m during glacial periods. A reasonable scenario is that species invaded caves during glacial periods when forests extended down to the altitude of the caves, then interglacial warming caused the extinction of surface populations of the species that invaded caves. Peck tentatively dates the time of isolation of the five arthropod species found in Grand Canyon caves. Two species, a mite (*Rhagidia*) and a dipluran (*Haplocampa*), are either conspecific with or most closely related to species in northern boreal forests. Based on their taxonomic relationships and the amount of modification, Peck tentatively concludes that they have been isolated since the Wisconsin glaciation. The leiodid beetle *Ptomaphagus cocytus* and the telemid spider *Telema* sp. are strictly troglobitic and were probably isolated in the Sangamon interglacial. Finally, the collembolan *Tomocerus* is eyeless and morphologically specialized, indicating isolation in the Yarmouth interglacial or earlier.

Peck (1981b) has proposed an even more detailed scenario for the time of invasion of the *Ptomaphagus hirtus* species group, using cladistic analysis along with new information on the timing of Pleistocene glaciations (Table 7–1, Fig. 7–1). The surface ancestors of the cave-limited species probably occurred in medium- or low-elevation mesic forests, as does the present-day *P. shapardi*. The basic scheme of cave invasion and isolation is as follows. Small allopatric populations were isolated in caves during the beginnings of interglacials, which by their drying and warming effects made surface habitats uninhabitable. Species isolated in one interglacial were probably capable of overland dispersal in the next glacial, which accounts for the repeated speciation within phyletic lines, as shown in Figure 7–1. The cladogram proposed by Peck will undoubtedly be modified as more information becomes available, especially about the number and timing of interglacials and about shared advanced (derived) characters. The importance of Peck's work is not in the accuracy of the cladogram but in the rigor and falsi-

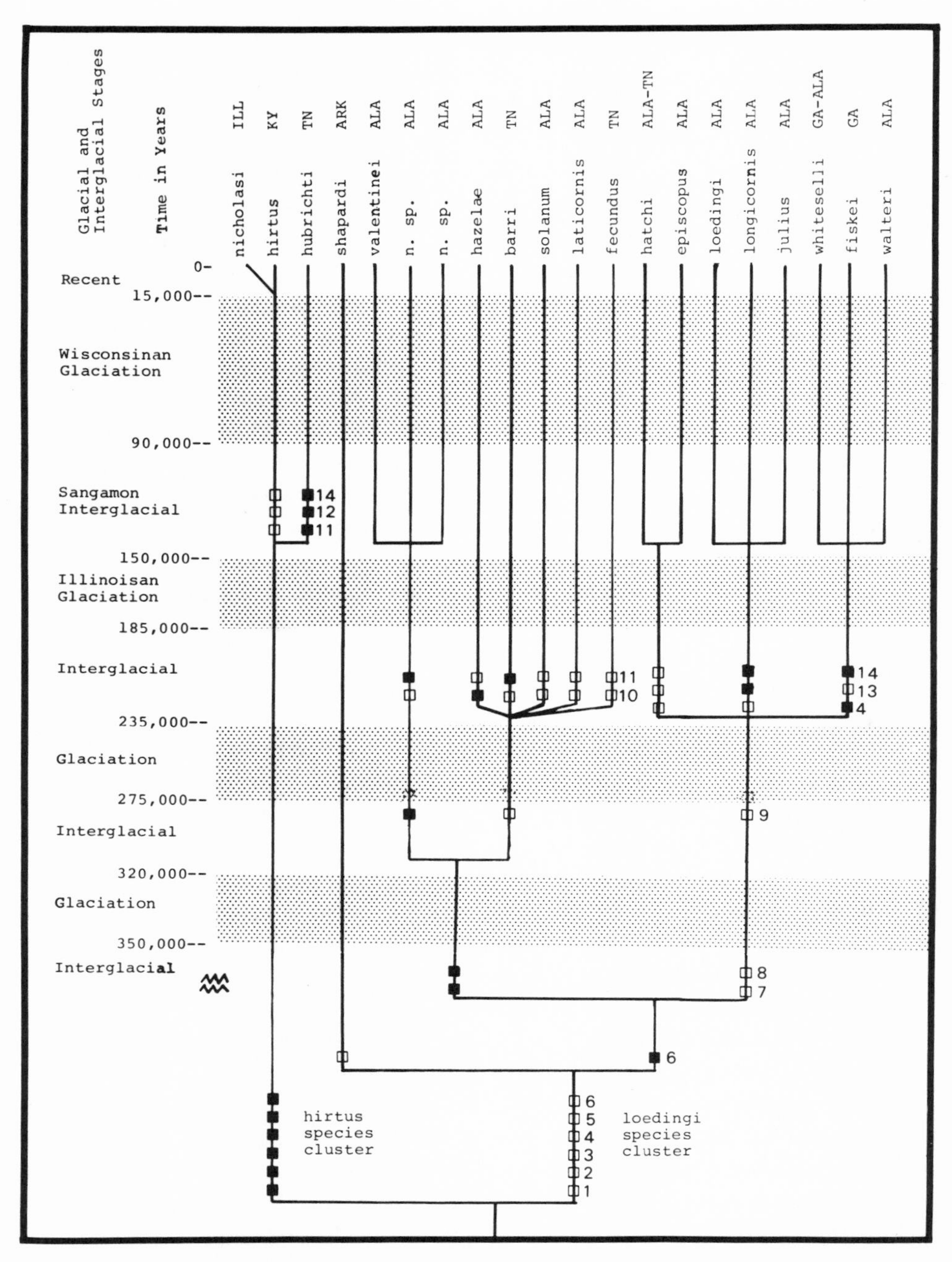

Figure 7–1 Cladogram of *Ptomaphagus hirtus* species group. Times of glacials and interglacials are derived from isotope studies of the time and temperature of deposition of cave stalagmites. Times of branching were determined in part by hybridization and allozyme studies. Lineage splits and character origin are indicated at the most recent likely times for the events. (From Peck 1981b.)

Table 7–1 Ancestral (plesiomorphic) and derived (apomorphic) characters of the *Ptomaphagus hirtus* species group, used in constructing the cladogram in Figure 7–1. Possible convergent characters are marked with an asterisk. (From Peck 1981b.)

		Character state	
Number	Character	Ancestral	Derived
1	Spermathecal shaft	Longer, thinner	Shorter, thicker
2	Spermathecal orifice	Posteriorly oriented	Laterally oriented
3	Aedeagus	Curved	Straight
4	Mesosternal carina*	Lower	Higher
5	Pronotal strigae	Present	Reduced or absent
6	Eyes*	Cluster of facets	Unpigmented spot
7	Spermathecal shaft	Thicker	Thinner
8	Posterior end of spermatheca	Not or slightly expanded	Greatly deflected
9	Spermathecal crest	Small	Large
10	Aedeagal tip	Straight	Upturned
11	Hind pronotal sides	Curved	Straight
12	Body size*	Smaller	Larger
13	Antennal segment III*	Subequal to II	Longer than II
14	Antennae*	Shorter	Longer

fiability of the hypothesis. The basis for the cladogram is the shared derived characters (synapomorphies) given in the table. As these change, the cladogram changes in an unambiguous way. Indeed, the main advantages of cladistic analysis are its lack of ambiguity and its utility in testing hypotheses (Nelson and Platnick 1981). As other cladograms are developed for other groups, the generality of Peck's scenario can be determined. At present, all that can be said is that such a scenario seems reasonable for terrestrial troglobites in areas affected by the Pleistocene, which includes the great majority of terrestrial troglobites.

There is, however, one situation where even the most general hypothesis about Pleistocene effects is in doubt. This concerns terrestrial troglobites in caves in the lowland tropics, where Pleistocene effects were not so severe. It may be that Pleistocene effects were sufficient for the isolation of some species in lowland caves (Reddell 1981), but this is disputed (Mateu 1980). Organisms undoubtedly enter caves for a variety of reasons, such as avoiding physiological stress or competitors or predators. The difficulty is in determining how isolation could result.

One possibility is that a species experiencing intense competition or predation and using caves as refugia may become isolated or nearly isolated in caves because of the biotic pressures.

Aquatic Species In contrast to terrestrial species, a much wider range of times of isolation in caves, from late Pleistocene to late Paleozoic, has been hypothesized for aquatic species. One reason for this diversity is that aquatic cave species have originated in two different ways. Some groups, such as North American cave crayfish, arose from freshwater species in surface streams that subsequently became extinct (Hobbs and Barr 1972). Other groups, including many amphipod species, arose from marine species.

Typical of the latter group is the amphipod genus *Stygobromus,* known only from subsurface habitats and with a Holarctic distribution. From the absence of any morphologically similar marine species, the extensive speciation in the genus (over 100 species in North America), and their Holarctic distribution, Holsinger (1978) suggests that the group was isolated in subsurface habitats prior to the breakup of Laurasia sometime in the late Paleozoic or early Mesozoic. Similar arguments have been made for other subsurface crustaceans (reviewed by Danielopol 1980). Some groups are thought to have been isolated by the recession of Miocene Seas, because of the close match between their present localities and the Miocene shoreline (Fig. 7–2). The original nonmarine habitat for *Stygobromus* may have been anchialine caves—caves, hollows, and depressions along coastal areas with mixohaline water (Sket 1981). Whether or not *Stygobromus* dates from the Paleozoic, it is certainly pre-Pleistocene. For aquatic cave species the main effect of the Pleistocene was extinction, with occasional populations surviving in caves beneath glacial ice (Holsinger 1981).

Isolation in subsurface waters is not the same as isolation in caves. Although there are a variety of subsurface habitats, which will be considered in more detail later in the chapter, it is convenient to distinguish interstitial and karstic habitats (Henry 1978); the former consist of tiny water-filled spaces in habitats such as the underflow of rivers, and the latter consist of cave waters and associated habitats. Of the forty-eight *Stygobromus* species from the eastern United States treated by Holsinger, thirty-eight are known predominantly from caves. But most of these apparently karstic species are known primarily from temporary drip pools and similar habitats, indicating that cave waters are not their primary habitat. Only nine species are primarily stream dwellers, and these are the genuine cave species. The colonization of

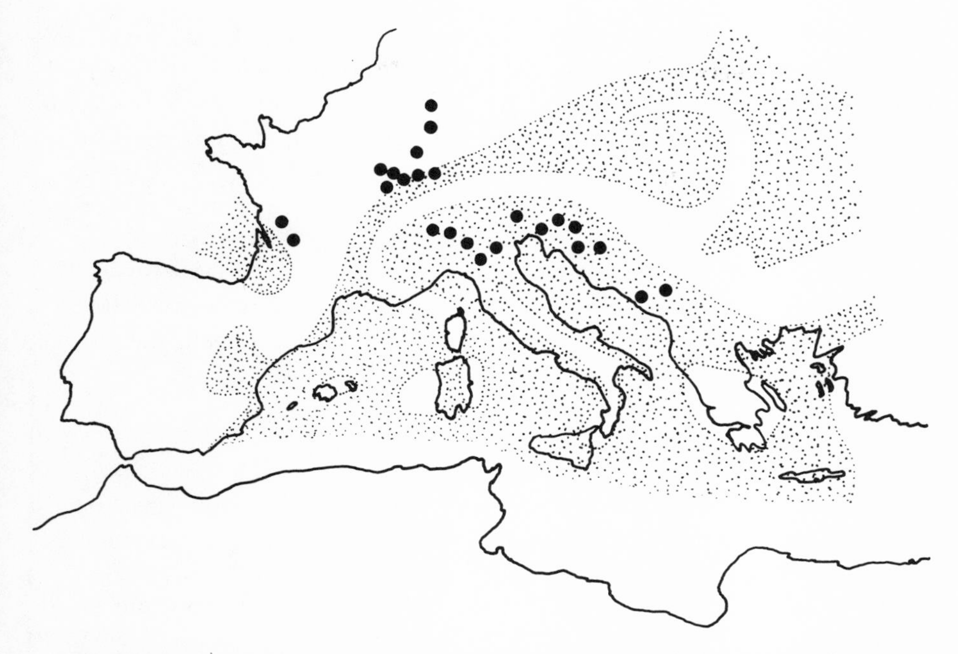

Figure 7–2 Distribution of cavernicolous isopods of the genera *Caecosphaeroma* and *Monolistra* (dots) superimposed on the extent of Miocene seas (shaded area). (Modified from Vandel 1964, by permission of Pergamon Press Ltd.)

karstic waters from interstitial waters and the colonization of cave waters from other karstic waters may be considerably more recent than the original isolation in phreatic waters. Our ignorance about the times of colonization of cave waters is complete.

By far the most thorough and convincing discussion of the age of a subsurface fauna is Danielopol's (1980) on the freshwater interstitial ostracods. Relying on the extensive fossil record and distributional data, Danielopol has pinpointed the earliest possible time of isolation. Although ostracods are primarily interstitial and for the most part accidental in caves, their history can serve as a model for groups of species in caves but without a fossil record. Danielopol concludes that the earliest freshwater interstitial ostracods—such groups as *Kovaleuskiella* in the Limnocytheridae—are early Oligocene in age and that isolation in fresh water occurred during marine shoreline regressions. Although ostracods and other crustacea are continually invading marine sands and anchialine caves (Sket 1981), the isolation caused by sea regression is required for the evolution of a subterranean fauna. Because of interstitial species that are common in the tropics, for example, *Darwinula*, Danielopol also concludes that the Pleistocene was not important in isolating ostracods.

Species that invaded cave streams from surface streams seem to have a more recent origin. For example, Hobbs and Barr (1972) suggest that the troglobitic crayfish in the *Orconectes pellucidus* group are remnants of a Miocene stock that were isolated in caves because of regional uplift in the late Pliocene or early Pleistocene. The Mexican cave fish *Astyanax mexicanus* is still more recent, having been isolated by stream capture relatively late in the Pleistocene (Mitchell, Russell, and Elliott 1977).

Dispersal

Studying the time of isolation of populations in caves is only the first part of the historical biogeography of cave faunas. A critical question is how much dispersal to other caves has occurred since the population was isolated. Put another way, does each cave species represent an independent invasion by a surface ancestor, or are groups of cave species the result of a single invasion with subsequent underground dispersal and speciation? Since the habitats where a species can potentially survive differ greatly for the terrestrial and aquatic faunas, they will be treated separately.

Terrestrial Species Many terrestrial species are often limited to one or several caves in a continuous outcrop of limestone, and species in caves separated by nonsoluble rocks such as shales and sandstones are often different. The classic case of species with small ranges largely defined by geologic barriers occurs in the carabid beetle genus *Pseudanophthalmus* in Appalachian caves (Barr 1967c). At least forty-eight described and undescribed species are found in caves in Virginia and northeastern Tennessee (Barr 1981); the ranges of twenty-seven are shown in Figure 7–3. Without exception, their ranges are within continuous outcrops of limestone. Assuming for the moment that each species represents a separate invasion by a surface ancestor, can the range of each be explained by underground movement through air-filled passages? One of the widest-ranging species is *P. delicatus* (A in Fig. 7–3) with a maximum linear extent of 45 km. Most of its range can be explained by underground movement through air-filled passages. There are over a hundred known caves within the range of *P. delicatus* (see Holsinger 1975). Even if all the accessible caves are known, there

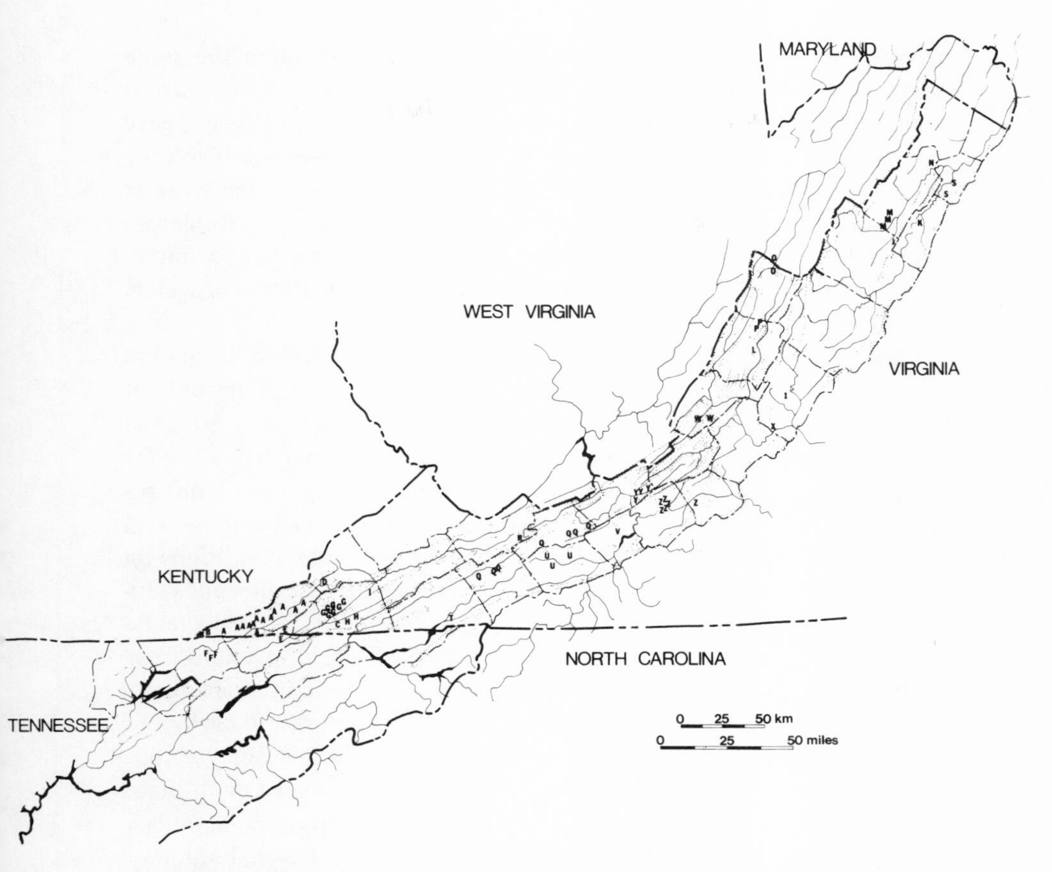

Figure 7–3 Ranges of all known species of *Pseudanophthalmus* in the *hirsutus, jonesi, hubbardi, petrunkevitchi,* and *pusio* groups in Virginia and adjoining areas of Tennessee. Major rivers and mountain ridges are shown on the map. Caves are developed along valleys or the flanks of ridges. The 27 species are indicated by capital letters and the number 1, as follows. *Hirsutus* group: A, *delicatus;* B, *hirsutus;* C, *sericeus. Jonesi* group: D, *cordicollis;* E, *longiceps;* F, *pallidus;* G, *seclusus;* H, *thomasi;* I, sp. 1. *Hubbardi* group: J, *avernus;* K, *hubbardi;* L, *intersectus;* M, *limicola;* N, *parvicollis;* O, *potomaca;* P, sp. 2. *Petrunkevitchi* group: Q, *hoffmani;* R, *hortulanus;* S, *petrunkevitchi;* T, sp. 3; U, sp. 4; V, sp. 5. *Pusio* group: W, *nelsoni;* X, *pontis;* Y, *punctatus;* Z, *pusio;* 1, sp. 6. (Map courtesy of Dr. John R. Holsinger.)

are probably several times that number of entranceless caves. Curl (1966) points out that the formation of an entrance is usually unrelated to formation of the cave and is essentially a random event whose probability of occurrence depends on the length of the cave. For a variety of karst areas, Curl estimates the fraction of entranceless caves at between 0.50 and 0.95, the median being 0.79. Although the large number of caves may explain much of the range of *P. delicatus,* it cannot explain its occurrence on both sides of the Powell River. There may well be caves beneath the Powell River (see below), but if so they are certainly water-filled. Barr and Peck (1965) believe, on the basis of both range data and their experiments, that cave beetles can tolerate several hours of water immersion. Thus direct cave-to-cave movement, usually in air-filled cavities but at least occasionally under water, can explain the ranges shown in Figure 7–3.

This pattern cannot be general. For example, the troglobitic spiders *Phanetta subterranea* and *Porrhomma cavernicolum* occur throughout the region shown in Figure 7–3. These "species" may be a group of sibling species or semispecies, or it may be that these spiders have far greater dispersal abilities than *Pseudanophthalmus* beetles. Until recently, it was not clear how such dispersal occurs, because in general troglobites are not able to survive the physical and biotic conditions on the surface. However, recent work by Juberthie and his colleagues (Juberthie, Delay, and Bouillon 1980; Juberthie and Delay 1981) has made it at least possible to explain such large ranges. They have found troglobites under the deepest layer of soil in mountains, in compartments consisting of cracks and fissures in the superficial part of the rock, and in interconnected spaces in scree and talus. Juberthie calls this habitat, which is most common in nonlimestone areas but can occur in limestone areas as well (Juberthie and Delay 1981), the "superficial underground compartment" (*milieu souterrain superficial*). This habitat may have enormous importance in dispersal of cave species because it makes habitable areas much more continuous (see Fig. 7–4). Its extent is unknown, and while most of the troglobite species known from nearby caves have been found in the superficial underground compartment, some groups appear to be relatively abundant (such as the leiodid beetle *Speonomus*) and others relatively scarce (such as the carabid beetle *Aphaenops*) compared to cave populations (Juberthie, Delay, and Bouillon 1980).

The range maps of *Pseudanophthalmus* shown in Figure 7–3 can also be reinterpreted in view of Juberthie's work. The assumption previously used was that each species represented a separate invasion by a surface species. By the assumption of a separate invasion, a vicariance viewpoint (see Nelson and Platnick 1981) was taken, at least with regard to movement between caves. A center of origin viewpoint, with subsequent underground dispersal and speciation, can be taken by assuming that each species group represents a single invasion. For example, one could hypothesize that the *P. hirsutus* group arose from a single invasion somewhere in the present range of *P. delicatus*. Infrequent dispersal, via caves and the superficial underground compartment, could allow populations to differentiate and give rise to *P. hirsutus* and *P. sericeus*. The two hypotheses can be separated when cladograms and more information about the superficial underground compartment are available.

Aquatic Species The potentially habitable areas are more complex for aquatic cave organisms than for terrestrial species. I will begin by considering the nature and continuity of subsurface water habitats, then consider if the various subsurface habitats are distinguished by the

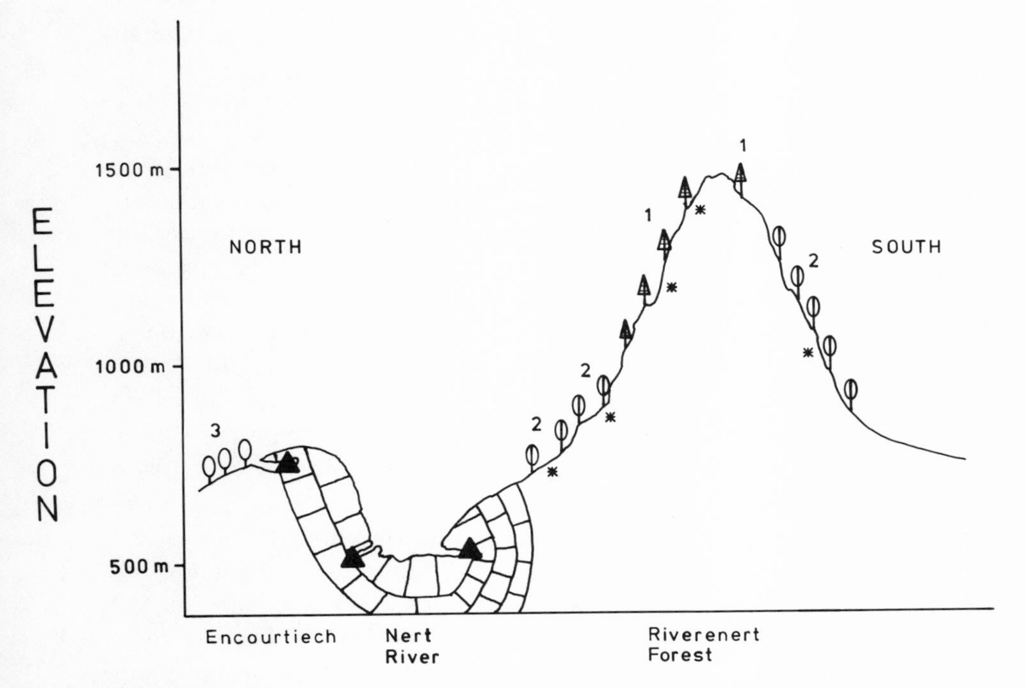

Figure 7–4 Schematic representation of the distribution of the beetle *Speonomus hydrophilus* near the Nert River in southern France. The triangles represent localities in caves and mines, and the asterisks represent localities in the superficial underground compartment. The forest types are 1, fir; 2, beech; and 3, oak. The blocks in the river valley indicate the extent of limestone. The horizontal distance represented is approximately 5 km. (Modified from Juberthie, Delay, and Bouillon 1980.)

organisms inhabiting them, and finally make some inferences about actual movement of organisms.

There is no single, universally accepted theory about the development of caves. Some basic questions that are of considerable interest to cave zoogeographers do not have completely satisfactory answers. For example, little is known about the rate of movement, if any, of water at great depth (Bögli 1980). Lack of a universal theory of cave development, although partly the result of the usual problems of different languages and national traditions, is also due to the fact that caves develop in various ways and that different countries often have different kinds of caves. For example, of the thirty caves in the world that are longer than 26 km, fourteen are in the United States and only three are in France. None of the twenty-seven caves deeper than 840 m is in the United States and eight are in France (Bögli 1980). In the following brief synopsis, I have followed Bögli's discussion of cave development in thick, relatively flat-bedded limestone, which is generally agreed on internationally.

In a karst aquifer, three vertical zones can be distinguished: (1) an upper, dry zone (inactive vadose); (2) a periodically flooded zone (active vadose); and a lower, continuously flooded zone (phreatic). In this case the word phreatic is used in a more restricted sense than before, meaning water below the water table. The water table itself is conceptual rather than real, defined as the level to which water will rise under hydrostatic pressure. Caves can develop wherever there is water available to dissolve limestone; they can form in the active vadose zone or at a depth of several thousand meters in the phreatic zone (Bögli 1980). However, it is in the shallow phreatic zone that cave formation is especially favored. This is supported by considerable empirical data on cave elevations, and there are geochemical reasons why cave formation is more likely near the water table. Deeper water is likely to become saturated with carbonate ions, but shallow phreatic water is more likely to remain active because of what Bögli terms mixing corrosion. Two bodies of water, both saturated with Ca^{+2}, but at different CO_2 pressures and hence different concentrations of Ca^{+2}, become corrosive (able to dissolve more $CaCO_3$) when mixed. Corrosion and mixing corrosion proceed at a very slow rate until capillary tubes or cracks reach a diameter of approximately 1 mm, at which point solution increases because of increased water velocity and turbulent flow. Consequently, in a mature karst area one would expect to find two relatively distinct tube sizes—one set of tiny tubes (about 1 mm in diameter) and another set of much larger tubes, or caves. Thus in the horizontal plane there would be two more or less distinct habitats. Caves may occur throughout the vertical plane but are likely to be concentrated at one or several levels corresponding to former levels of the water table.

If the same organisms could live in all subsurface water in a karst area, their habitat would be highly connected, with little physical barrier to movement. This is not the case, but before considering the biological data, we need to review the subsurface habitats outside of soluble rock. There is a nearly inpenetrable jungle of terms for noncave subsurface habitats. I follow Henry's (1978) classification here for both karstic and nonkarstic habitats, but there are many others. The systematic study of shallow subsurface freshwater habitats was made possible in large part by the invention of the Bou-Rouch pump, a device that can extract water from a few centimeters to a few meters below the surface (Bou and Rouch 1967). Henry recognizes three major habitats, which he terms interstitial to differentiate them from the karstic habitats:

1. *Nappes phreatiques,* the gravels along the sides of streams and rivers.
2. *Nappes fluviales,* the underflow of streams and rivers.
3. *Nappes perchées,* subsurface habitats above the water table.

These, together with karstic habitats, are shown in Figure 7–5.

Although the fauna of the interstitial habitats shown in the figure have been much less thoroughly collected than cave faunas, and the phreatic zone of karst areas is rarely accessible, the following picture is emerging. First, the interstitial fauna is relatively distinct from the karst fauna. There are several examples of closely related species or subspecies occurring in one but not both major habitats. Magniez (1976) provides an exceptionally clear one. In the Lachein Valley near Moulis, France, the isopod *Stenasellus virei hussoni* is known only from caves, and *S. virei boui* is known only from interstitial habitats, especially the nappes fluviales. In the eastern United States, the amphipod *Stygobromus pseudospinosus* is known only from a cave in Virginia, and its primary habitat is likely to be tiny tubes and cracks in the active and inactive vadose zone in karst. The closely related *S. spinosus* is interstitial, known from seeps and small springs similar to the nappes perchées (Holsinger 1978).

Second, if the subterranean fauna is very rich, the different karstic zones tend to have different faunas. The isopod *Proasellus valdensis* is primarily found in streams in the active vadose zone (Henry 1978), as are many North American isopods in the genus *Caecidotea*. Phreatic cave waters are not well known, but at least in Texas the fauna from ar-

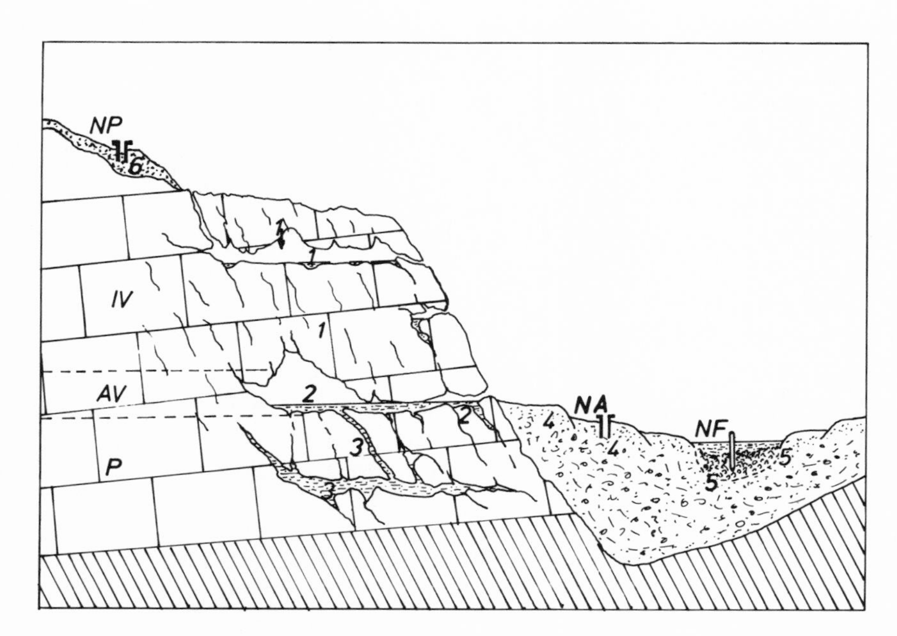

Figure 7–5 Major subsurface habitats. Karstic: 1, inactive vadose zone; 2, active vadose zone; and 3, phreatic zone. Interstitial: 4, nappes phreatiques; 5, nappes fluviales; and 6, nappes perchées. (After Henry 1978.)

tesian wells that pump phreatic water is distinct from fauna in caves in the vadose zone (see Holsinger and Longley 1980). Third, drop pools and similar habitats in the inactive vadose zone are rarely if ever primary habitats for organisms. Drip pools have long been favorite collecting spots because they are easily sampled, but populations there are usually small and rarely persistent (Henry 1978, Holsinger 1978). Henry suggests that tiny tubes and cracks in the inactive vadose zone may often be the primary habitat for these populations. Fourth, there can be considerable movement through and among the different habitats, as shown by the frequent occurrence of organisms in drip pools. Fifth, in areas with a depauperate subterranean fauna, one species may occupy both karstic and interstitial habitats. On the right bank of the Rhone River, the amphipod *Niphargus virei* occurs in the active vadose zone, but on the left bank it is absent. In its place is *N. rhenorhodanensis,* which is found only in interstitial habitats on the right bank (Turquin 1981). To summarize, organisms seem to distinguish between the habitats shown in Figure 7–5, but movement through nonpreferred habitats is at least possible. Unfortunately, the quantitative importance of movement is not known, but some inferences can be made from range data.

Compared to *Pseudanophthalmus* beetles in Virginia and eastern Tennessee (Fig. 7–3), no group of aquatic organisms shows a similar pattern of high endemism and, by implication, highly restricted dispersal. The closest approach to the *Pseudanophthalmus* pattern is the amphipod genus *Stygobromus,* with seventeen species in the area covered by Figure 7–3, many of which are known from only one or two localities (Holsinger 1978). But at most four of these species are genuinely cave rather than interstitial species. The small known ranges of most *Stygobromus* either result from inadequate collecting in interstitial habitats or from extensive speciation of the interstitial species with low rates of dispersal.

Besides the tendency toward larger ranges than terrestrial species, aquatic species tend to occupy a greater fraction of the caves within their range, which also indicates higher rates of dispersal. Both *Caecidotea* isopods and *Crangonyx* amphipods occupy a greater number of caves than *Pseudanophthalmus* in Lee County, Virginia (Table 7–2). It is most likely that the major habitats for both *Caecidotea* and *Crangonyx* are streams in the active vadose zone and that the other subterranean habitats shown in Figure 7–5 are primarily dispersal routes. The apparent higher movement rates for aquatic species are probably caused by several factors, including the large number of caves connected by water-filled passages and the horizontal movement of water, especially during flooding.

In spite of the higher rates of movement and the possibility of interstitial habitats as dispersal routes, not all aquatic cave species have large ranges. One aquatic species with a range like that of *Pseudanophthalmus* is the isopod *Lirceus usdagalun* in Lee County, Virginia (Table 7–2). Known from only four caves, its small range is a consequence of intense competition from *Caecidotea* and *Crangonyx* (see chapter 6) and its relatively recent isolation in caves.

If, instead of comparing aquatic and terrestrial cave species, we compare aquatic cave and aquatic surface species, a different pattern emerges. There is a relatively rich amphipod fauna in streams in glaciated eastern North America. But aquatic fauna in caves in glaciated regions is almost nonexistent. In Holsinger's (1981) review of those subterranean amphipods that occur north of the southern limit of Pleistocene glaciation, most if not all of the eleven species are probably interstitial rather than karstic. He provides convincing evidence that at least four of these species survived glacial periods in deep groundwater

Table 7–2 Comparison of distributional characteristics of troglobitic beetles (*Pseudanophthalmus*), isopods (*Caecidotea* and *Lirceus*), and amphipods (*Crangonyx*) found in Lee County, Virginia. (Data from a review of Virginia cave fauna by Holsinger and Culver, in preparation.)

Group	No. cave records	No. species	No. county endemics
Pseudanophthalmus	21	6	4
Crangonyx	42	1	0
Caecidotea	33	3	1
Lirceus	4	1	1

beneath the ice. Especially convincing is the case of *Stygobromus canadensis* in a cave that is still partly covered by a glacier. Only *Bactrurus mucronatus* has moved more than 150 km north of the glacial maximum. That species' movement is not surprising since it is found in tile drains on the edge of cultivated fields in glacial drift plains (Holsinger 1981), an almost ideal habitat for rapid migration.

Caves as Islands

The analogy between caves and islands has been widely applied to both evolutionary time (Barr 1967b) and ecological time (Culver 1970b, 1971b). In a literal sense, for caves to be islands means that the only habitable areas are the caves themselves, with at least occasional dispersal between them. This cannot strictly be true, because some terrestrial troglobites can survive and reproduce in superficial underground compartments (see Fig. 7–4), and some aquatic troglobites can survive and reproduce in interstitial habitats (see Fig. 7–5). If these habitats themselves are not continuous, they can be thought of as additional islands. But the real point of the caves-as-islands analogy is that some biological processes are the same in both. In particular, for the cave–island analogy to hold in evolutionary time, populations in caves must be sufficiently isolated gene pools that allopatric speciation occurs. For the analogy to hold in ecological time, the number of species in a cave must be determined by a balance between immigration and extinction (MacArthur and Wilson 1967). This section will emphasize islandlike processes in ecological time, and the concluding section will consider evolutionary questions.

Judging by the criticisms, which range from profound (Simberloff 1976, Connor and McCoy, 1979) to churlish (Gilbert 1980), it is clear that MacArthur and Wilson's theory has not been satisfactorily demonstrated when applied to virtual islands such as caves. My own and others' incautious enthusiams for island biogeography theory has resulted in incautious criticism, as exemplified by Gilbert's restriction of the theory to situations where species interactions are unimportant and only the simplest stochastic processes are important.

What is needed to demonstrate MacArthur and Wilson's theory that an equilibrium between immigration and extinction determines species numbers? The easiest and most commonly used evidence is that z in the following equation is approximately 0.26:

$$S = CA^z \qquad (7\text{-}1)$$

where S is species numbers, A is area, and C and z are fitted constants. But this is neither necessary nor sufficient to demonstrate a dynamic equilibrium. Not all equilibrial systems have z-values near 0.26 (May 1975), and a z-value near 0.26 does not necessarily imply an equilibrial system (Connor and McCoy 1979). A more important set of observations concerns changes in species composition, without which MacArthur and Wilson's theory does not hold. Simberloff (1976) points out that the measurement of ''unimportant'' transients, such as birds feeding on a fruit, can give rise to spuriously high values of immigration and extinction. Crawford (1981) takes the extreme view that the entire aquatic fauna in caves is in a suboptimal habitat and is in transit. The discussion in the previous section indicates that this is most unlikely.

There are, however, microgeographic scales in which Crawford's contention is approximately right. For many aquatic species living in streams, the rocks in a riffle are analogous to fruit trees visited by birds. Most immigrants from one rock to another do not reproduce, and most, but not all, extinctions are emigrations. To consider the individual rocks in a riffle as islands is clearly inappropriate in island biogeography theory. Some proportion, probably a significant proportion, of recruitment into the population is by immigration rather than births. But this system of ''unimportant transients'' is the basis of species interactions (see chapter 6), and the entire population is in this system of patches. On this scale the islandlike characteristics are the patchiness and the risk in moving from patch to patch.

Different riffles in the same stream are also patches of habitable space separated by areas of increased risk (pools), where food is scarcer and predators more common. It is possible that riffles are islands in the sense of MacArthur and Wilson. They do share several characteristics with islands. Even when a species is alone in a stream, it does not occur in all riffles. For example, the very large population of *Gammarus minus* in Benedict's Cave occurs in only 88 percent of the riffles ($n = 73$), and most populations occupy a considerably smaller fraction (Culver 1973a). This is consistent with the view that riffles are islands in an archipelago since the proportion of riffles (islands) occupied ($\hat{p}$) should be

$$\hat{p} = 1 - \frac{e_i}{m_i T} \tag{7-2}$$

where e_i is the riffle extinction rate, m_i is the migration rate from one riffle to another, and T is the total number of riffles (Culver 1971b).

The second point of agreement with island biogeography theory is that there is turnover in species composition in riffles. There are no quantitative data on this, but repeated sampling in several caves in the Appalachians has turned up repeated cases of change in species composition. There are, unfortunately, no data on area effect, which would be especially difficult to explain, because riffles in a stream of constant width tend to be similar in size (Leopold, Wolman, and Miller, 1964). There are also strong indications that while a cave stream is a patchy environment, the scale is smaller than that of an island in the sense of MacArthur and Wilson. Organisms in a riffle have a high washout rate (see Fig. 6–6), resulting in considerable mortality but also immigration. This suggests that emigration plays an important role in population recruitment, indicating that the scale is too small to consider riffles as islands. The frequent occurrence of species in drip pools indicates considerable movement between cave streams. That is, there is migration from other archipelagoes. If this movement were very infrequent, one would expect single-cave endemics, but such cases are rare, and many of these are interstitial species that only occasionally enter caves (see Holsinger 1978).

Patchiness on a small scale also occurs in some terrestrial cave communities Those species that specialize on vertebrate (but not bat) guano exist in a highly patchy environment because a mound of dung is a patch of habitable space that is slowly used up. The short duration of a patch precludes a direct analogy with island biogeography theory, but the system has many similarities with patches of the rocky intertidal cleared by starfish predators and slowly overgrown by mussels and other species (Paine and Levin 1981).

From the point of view of the physical structure of the environment and the distribution of organisms, the natural scale at which an island analogy for terrestrial cave fauna might exist is that of the cave itself. Many caves end in siphons or in inpenetrable breakdown, both of which mark the approximate end of the terrestrial cave environment. The area of a cave may be correlated with population size and, if equilibrium island biogeography theory holds, should be inversely correlated with extinction rate. There has been, unfortunately, no attempt to examine the applicability of island biogeography theory to the terrestrial cave fauna except for Vuilleumier's (1973) analysis of the terrestrial and aquatic fauna in Swiss caves. He examined only area effect and failed to find any for the troglobitic fauna. In the next section, an analogy between whole karst areas and islands will be examined, but the time scale is likely to be evolutionary rather than ecological time.

For the aquatic fauna, two scales have some potential for an island analogy. The first scale is cave streams or, more precisely, sections of cave streams, which roughly correspond to separate caves, separated by waterfalls or long pools. A small cave usually has one stream that may siphon into a large pool at the end of the cave, with other streams continuing in passages too small to follow. A large cave system usually has several distinct streams and stream segments. This confusion is simplified in practice because most biological collections from a cave are from a stream segment. The second scale is the drainage basin, that is, all the karst water that resurges at the same spring.

Several bits of evidence support the analogy between cave stream segments and islands. First, cave species are not found in all caves

Table 7–3 Changes in species composition in Upper Martha's Cave, Pocahontas Co., W.Va. An X indicates the species was present, and a dash indicates absence. (Data from Culver 1970b.)

Date	*Stygobromus emarginatus*	*Stygobromus spinatus*	*Gammarus minus*
1/67	X	X	—
6/67	—	X	—
8/67	—	X	—
11/67	X	X	—
3/68	—	—	X
8/68	X	X	X
10/68	—	—	X
2/69	—	—	X

Table 7–4 Total cave amphipod and isopod species, mean number of species per cave, and maximum number per cave. (From Culver 1976; reprinted by permission of the University of Chicago Press, © 1976, the University of Chicago.)

Karst area	No. caves	Total no. species	Max./ cave	Mean/ cave	S.E.
Powell Valley, Va.	10	4	3	2.1	0.1
Upper Clinch Valley, Va.	10	7	4	2.1	0.3
Greenbrier Valley, W. Va.	10	6	3	2.5	0.4
Monongahela Valley, W. Va.	6	5	3	1.9	0.7
Indian Creek, W. Va.	4	4	3	1.8	0.6

within their range (see equation 7-2). Second, there is considerable species turnover. The extreme example shown in Table 7–3 represents at least some transients. Obviously nontransient invasions and extinctions have also been observed, including an invasion of the isopod *Lirceus usdagalun* in Gallohan Cave No. 2, Virginia, and the amphipod *Stygobromus spinatus* in Martha's Cave, West Virginia (Culver 1976). Third, the mean number of species per cave stream is lower than the maximum, usually three, allowed by competition (see chapter 6), thus indicating island effects. But the total number of species in a continuous karst area with many caves is higher, because the islandlike nature of the system allows for alternate stable communities (Table 7–4).

The major negative evidence for the caves-as-islands analogy is the lack of area effect (Culver 1970b, Vuilleumier 1973). Although Connor and McCoy (1979) point out that area effect has been overemphasized as demonstrating equilibrium island theory, Crawford (1981) claims that the lack of area effect indicates that island theory does not hold for caves. This calls for a brief consideration of the model of area effect for equilibrium island biogeography theory. In the simplest situation, a series of identical islands that are equally distant from a source area, the following equation should hold (MacArthur and Wilson, 1967):

$$\frac{dS}{dt} = m(P - S) - eS \qquad (7\text{-}3)$$

where S is the number of species on an island, P the species pool, m the migration rate, and e the extinction rate. A similar equation holds for movement within an archipelago, but with a more complex migration term. At equilibrium

$$\hat{S} = \frac{mP}{m + e} \tag{7-4}$$

The area effect is

$$\frac{d\ \ln\hat{S}}{d\ \ln A} = \frac{1}{\hat{S}} \cdot \frac{d\hat{S}}{d\ \ln A} = \frac{m + e}{mP} \cdot \frac{d\hat{S}}{de} \cdot \frac{de}{d\ \ln A} = \frac{-de/d\ \ln A}{m + e} \tag{7-5}$$

When m, the migration rate is large, the area effect is small. If $de/d\ \ln A$ is constant, then, for a given value of m, the area effect will be least when e is large. The lack of significant area effect may result from high m and e or from a failure of the equilibrium model to hold for aquatic cave faunas. Area effect will be considered in more detail in the next section.

No work has been done on applying island biogeography theory to separate drainage basins, and it would be very difficult to do more than examine area effect in an area this size. Migrations and extinctions in such an area would be nearly impossible to detect because the entire drainage would not be accessible.

History versus Equilibrium

The application of island biogeography theory to cave faunas has usually been on a smaller geographic scale than that used by historical biogeographers, but island theory can be applied at the level of karst regions, that is, on the scale of historical biogeography. Although an equilibrium hypothesis and a historical hypothesis may in some ways complement each other, they are distinct. Under the equilibrium hypothesis there is a balance between immigration and extinction either in ecological or evolutionary time. If the time scale is evolutionary, the immigrations are the isolation of species in caves. By contrast, under a historical hypothesis, the number of species is not at an equilibrium but is rather the unique result of the history of the area. For biogeographers of the center-of-origin school, the important parameters are distance from the group's center of origin as well as the number of isolating events.

In practice, the two hypotheses are difficult to separate. Since what I call the historical hypothesis is a series of hypotheses depending on, among other things, how much subsurface dispersal is hypothesized, a reasonable way to proceed is to attempt to falsify the equilibrium hypothesis, which is less ambiguous. If separate karst areas, often with hundreds of caves, are viewed as potential islands, it is impossible to measure species turnover even if it is occurring in ecological time. One would have to visit many caves repeatedly, and one could never be sure that what appeared to be migrations and extinctions were not simply movements within the karst islands, especially those involving entranceless caves.

The one thing that can be measured is area effect. The biology and statistics of area effect are very controversial (see May 1975; Connor and McCoy 1979; Sugihara 1981; Stewart 1981), but the following points seem to be emerging. First, there is no necessary correspondence between the equilibrium theory of island biogeography (see equation 7-5), and a z-value of approximately 0.26 in equation 7-1. A correspondence would require the additional assumption that the abundance of many species is lognormally distributed (Connor and McCoy 1979). Other species abundance curves, such as the broken-stick distribution, result in different z-values (May 1975). Second, z-values for flora and fauna from oceanic islands do tend to cluster between 0.2 and 0.3, while z-values for contiguous continental areas tend to be lower (MacArthur and Wilson 1967). Third, this clustering of z-values for islands may be a statistical artifact (Connor and McCoy 1979), although this has been disputed (Siguhara 1981). Fourth, the correct starting point for an analysis of area effect is to consider the untransformed linear model

$$S = C + zA \qquad (7\text{-}6)$$

which corresponds to species numbers being controlled by passive sampling from the species pool, with larger areas receiving effectively larger samples than smaller ones (Connor and McCoy 1979).

Even a simple species list for a karst region is difficult to obtain since it requires visits to many caves to adequately sample species, some of which are extremely rare. A second requirement for a useful analysis of area effect is that the karst regions be well defined and, ideally, that they differ only in area. The second requirement is more restrictive than it may sound. Many limestones, especially dolomitic limestones, show little karst or cave development (Bögli 1980), and it would be difficult to assess areas without also having a measure of cave development. The only analysis of this type that has been attempted is that of Culver, Holsinger, and Baroody (1973) for caves in the upper Kanawha valley of West Virginia. Since the publication of that paper, more complete species lists and more accurate measurement of karst areas have become available. Area effect in the upper Kanawha valley will be considered in some detail because it is the only such study.

The study area is shown in Figure 7–6. With one exception, karst regions are delimited by (1) sandstones and shales; (2) the Greenbrier

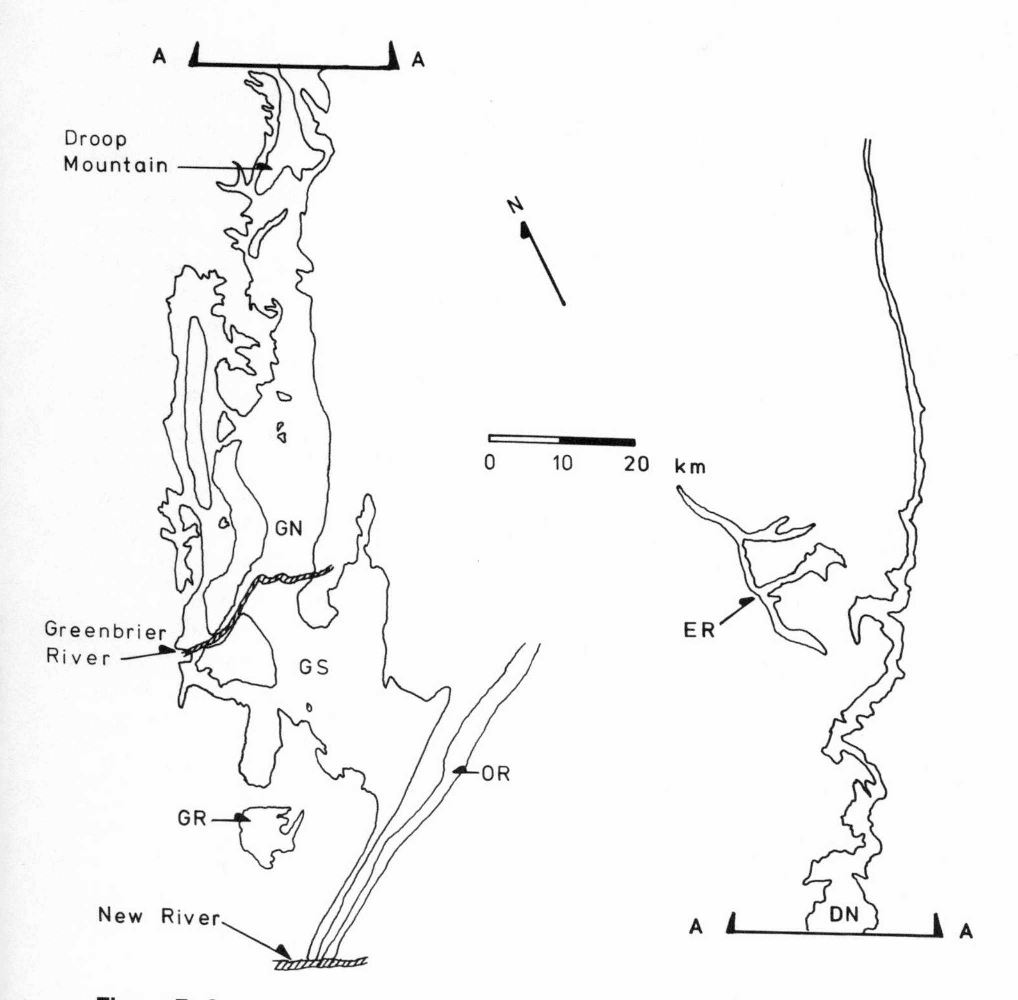

Figure 7–6 Karst areas of the upper Kanawha drainage, West Virginia. From south to north they are: GR, Greenville karst area (32 km²); OR, Ordovician limestone (58 km²); GS, Greenbrier Valley South (276 km²); GN, Greenbrier Valley North (360 km²); DN, Droop Mountain North (149 km²); and ER, Elk River karst (24 km²). (Modified from Culver, Holsinger, and Baroody 1973.)

River and New River, which cut through the Mississippian limestones; and (3) major drainage divides. The one exception involves the areas called Greenbrier North and Droop Mountain North, which are nearly but not completely separated by Droop Mountain. Few caves are known from the area immediately east of Droop Mountain that connects the two areas. More details can be found in Culver et al. (1973) and Holsinger et al. (1975). Two areas are not included. Narrow bands of limestone extend west of the New River in Mercer County, West Virginia, and these karst islands have not been thoroughly collected, nor even thoroughly explored for caves. Another karst island in Devonian limestone to the east of Greenbrier North, although included in the original analysis, has been eliminated because (1) the extent of the outcrop of Devonian limestone is uncertain, (2) caves are scarce, and (3) it has not been adequately sampled. To minimize the potential for showing a spurious area due to differences in sampling intensity (Simberloff 1976), it has been deleted from the present analysis. All other areas have high cave densities and have been intensively collected.

A complete list of the terrestrial troglobites for these areas is given in Table 7–5. Some of the species are poorly known, such as spiders and mites, and future taxonomic work is unlikely to change species numbers, but it may well raise the level of endemism. At present, 54 percent of the terrestrial species are thought to be endemic to the area. Following the suggestion of Connor and McCoy (1979) we examined area effect for four models: $S = C + zA$ (passive sampling); $\log S = C + zA$; $S = \log C + z \log A$ (exponential model); and $\log S = \log C + z \log A$ (power function). If the equilibrium model holds, one would expect the power function equation, or perhaps the exponential equation, to provide a better fit to the data than the passive sampling equation. Judged on the basis of lack of fit, only the exponential model fails (Table 7–6), and there is little difference in correlation of the other three models, although the linear model has a slightly higher correlation. There is one other indication that the equilibrium model does not hold. The exponent of the power function, $S = CA^z$, is 0.46, which is higher than usually reported for oceanic islands. Although that is not inconsistent with an equilibrium model, it does indicate low rates of migration and extinction (Culver, Holsinger and Baroody 1973), which in turn indicates that it would take a long time to reach equilibrium. The very strong area effect is consistent with the concept of completely isolated islands, whose populations are slowly going extinct (see Brown 1971). The pattern is not consistent with data collected from continuous continental areas. I thus conclude that karst areas are islandlike but not in equilibrium.

Table 7–5 Terrestrial troglobites of karst areas in the upper Kanawha River drainage. Areas are those indicated in Fig. 7–7. (Records are from Holsinger, Baroody, and Culver 1976.)

Class	Order	Species	Areas						Endemic?
			ER	DN	GN	GS	OR	GR	
Diplopoda	Chordeumida	*Pseudotremia fulgida*		X	X				Yes
		Trichopetalum packardi				X	X		No
		Trichopetalum weyeriensis	X		X	X			No
Insecta	Diplura	*Litocampa fieldingi*		X	X	X			Yes
	Collembola	*Pseudosinella gisini*	X	X	X	X			Yes
		Sinella hoffmani		X	X			X	No
		Sinella agna	X						No
	Coleoptera	*Pseudanophthalmus fuscus*		X	X	X	X		Yes
		Pseudanophthalmus grandis		X	X	X	X	X	Yes
		Pseudanophthalmus higginbothami		X	X				Yes
		Pseudanophthalmus hypertrichosis	X	X	X	X			No
		Pseudanophthalmus lallemanti			X				Yes
		Pseudanophthalmus subaequalis			X				Yes
		Pseudanophthalmus sp.					X		No
Arachnida	Acarina	*Rhagidia* sp.	X	X	X	X	X	X	No
	Pseudoscorpionida	*Kleptochthonius henroti*		X	X				Yes
		Kleptochthonius orpheus					X		Yes
		Kleptochthonius proserpinae			X				Yes
		Kleptochthonius hetricki				X			Yes
		Chitrella regina			X				Yes
	Araneae	*Anthrobia mammouthia*		X	X	X			No
		Bathyphantes weyeri			X	X			No
		Phanetla subterranea		X	X	X	X	X	No
		Porrhomma cavernicolum	X			X	X		No
Total (N = 24)			6	12	18	13	8	4	13

Table 7–6 Comparison of least-squares regression analysis for four equations of area effect of terrestrial cave species in the upper Kanawha valley: (1) $S = C + zA$; (2) $\log S = C + zA$; (3) $S = \log C + z \log A$; and (4) $\log S = z \log A + \log C$, where z and C are fitted constants. Type II regression (reduced-major-axis) parameters can be calculated as follows: RMA slope = least-squares slope/correlation coefficient, and RMA intercept = mean number of species − (RMA slope × mean area). p is the probability of a nonzero slope. (From Connor and McCoy 1979.)

	Regression model			
	1	2	3	4
Slope	0.035	0.0015	9.95	0.46
Intercept	4.9	0.72	9.43	0.054
Correlation	0.956	0.899	0.951	0.937
Lack of fit?	No	No	Yes	No
p	$>.99$	$>.99$	$>.99$	$>.99$

Examination of the residuals of the regression equations of Table 7–6 provides some support for Barr's hypothesis (1981) that beetles, and by implication other terrestrial troglobites, spread out from the Alleghenies during glacial periods and were isolated in caves during interglacials. If he is correct, then the more northern karst regions that are closer to the heart of the Alleghenies should have more species than expected from area effect alone. Examination of the residuals of the regression of untransformed variables (Fig. 7–7) indicates that the northern areas (ER, DN, and GN) do tend to have more species than expected, while southern areas have few, the only exception being the southern Ordovician area, which has more than the expected number of species. The same qualitative result obtains from the analysis of the other three regression equations in Table 7–6.

Lists of the aquatic species for these areas are given in Table 7–7. Species known only from deep gravels and mud (lumbriculid worms) or drip pools (some *Stygobromus* amphipods) are almost certainly interstitial rather than karstic and are not included in the analysis. However, their distributions are given in the table. Use of the same four regression equations as for the terrestrial analysis gave similar results (Table 7–8). Once again, the untransformed regression equation gave a slightly better fit. Contrary to the earlier work based on less complete data (Culver, Holsinger and Baroody 1973), there was a significant area effect for all four models, and z for the power function was 0.30, within the range reported for the biota of oceanic islands. The better fit of the

Table 7–7 Aquatic cave-limited species of karst areas in the upper Kanawha River drainage. Areas are those indicated in Figure 7–7. Interstitial species known from caves are listed at the end. (From Holsinger, Baroody and Culver 1976, except as indicated.)

Class	Order	Species	ER	DN	GN	GS	OR	GR	Endemic?
Cave species									
Turbellaria	Tricladida	*Macrocotyla hoffmasteri*		X	X	X			No
Gastropoda	Mesogastropoda	*Fontigens tartarea*				X	X	X	No
		Fontigens turritella[1]			X				Yes
		Fontigens holsingeri[1]		X					No
		Fontigens nickliniana[1]				X			No
Crustacea	Amphipoda	*Stygobromus emarginatus*	X	X	X	X	X		No
		Stygobromus spinatus		X	X	X	X	X	Yes
		Crangonyx sp.			X	X		X	No
	Isopoda	*Caecidotea holsingeri*	X	X	X	X	X	X	No
		Caecidotea scrupulosa[2]			X	X		X	No
	Decapoda	*Cambarus nerterius*	X		X				Yes
Total (N = 11)			3	5	8	8	4	5	3
Interstitial species									
Clitellata	Lumbriculida	*Stylodrilus beattiei*		X	X				Yes
		Trichodrilus culveri		X					Yes
Crustacea	Amphipoda	*Stygobromus mackini*[3]						X	No
		Stygobromus nanus[3]		X					Yes
		Stygobromus parvus[3]		X					Yes
		Stygobromus pollostus[3]			X	X			Yes
		Stygobromus redactus[3]					X		Yes

1. From Hubricht (1976).
2. Known from surface habitats in the Georgia. The species needs reexamination.
3. From Holsinger (1978).

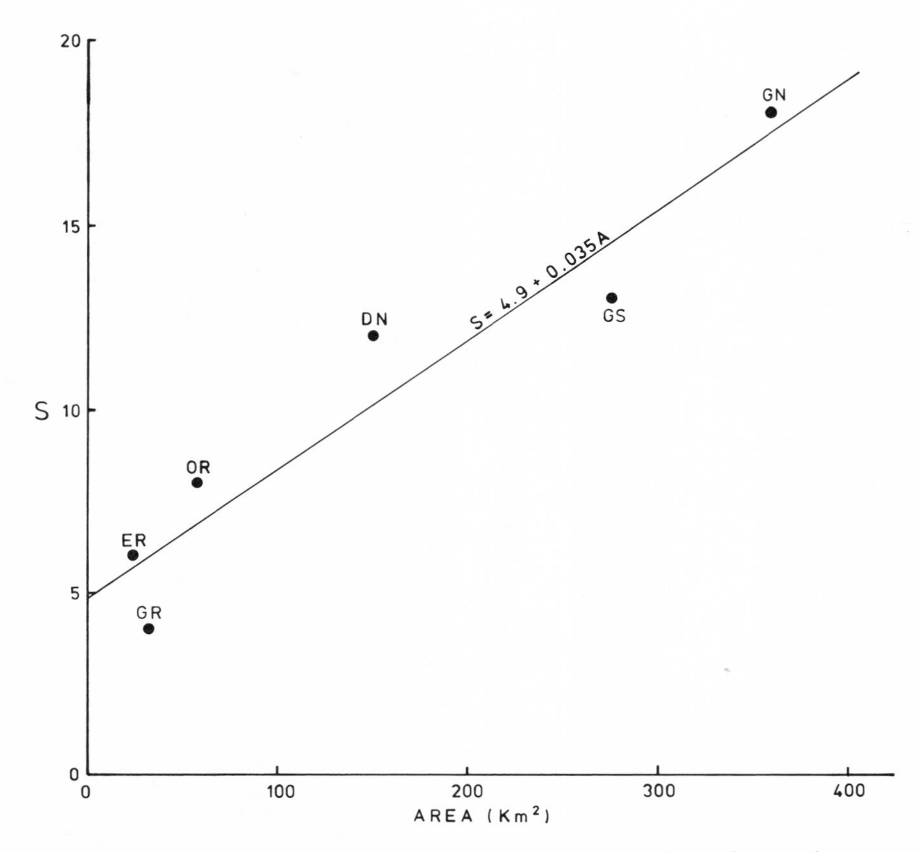

Figure 7–7 Linear least-squares regression of terrestrial species number on area for untransformed variables, $S = C + zA$. Labels of points are the same as in Figure 7–6.

untransformed model indicates that for aquatic species there is no reason to reject the passive sampling hypothesis. But the lower z-value of the power function, compared to that of terrestrial species, is consistent with higher migration rates and a faster approach to equilibrium. This is also borne out by the low level of endemism (see Table 7–7).

Summary and Prospect

Although historical biogeography has long been an important concern of cave biologists, we are at the threshold of a quantum jump in our understanding of historical cave biogeography. Perhaps the most important factor in this resurgence of biogeographic speculation lies in

Table 7–8 Comparison of least-squares regression analysis for four equations of area effect of aquatic species in the upper Kanawha valley: (1) $S = C + zA$; (2) $\log S = C + zA$; (3) $S = \log C + z \log A$; and (4) $\log S = z \log A + \log C$, where z and C are fitted constants. Type II regression (reduced-major-axis) parameters can be calculated as follows: RMA slope = least-squares slope/correlation coefficient, and RMA intercept = mean number of species − (RMA slope × mean area). p is the probability of a nonzero slope. (From Connor and McCoy 1979.)

	Regression model			
	1	2	3	4
Slope	0.014	0.0025	1.63	0.30
Intercept	3.44	1.27	1.90	0.27
Correlation	0.931	0.897	0.892	0.889
Lack of fit?	No	No	No	No
p	$>.99$	$>.99$	$>.99$	$>.99$

the changes that the entire discipline of historical biogeography has undergone. While both cladistics and vicariance are very controversial concepts, their proponents (see Nelson and Platnick 1981) have made considerable progress in formulating testable hypotheses and producing unambiguous rules for generating phyletic and biogeographic histories. The increased rigor and testability can be seen in the work of Danielopol (1980) and Peck (1981b), neither of whom hold vicariance views. But there are also several factors unique to cave biogeography that are important in the renaissance of biogeographic speculation. One is the possibility of dating the timing of Pleistocene interglacials and the age of caves using radioactive dating techniques (see Thompson, Schwartz, and Ford 1974). Another is the work of French cave biologists on noncave subsurface habitats for both terrestrial and aquatic species, work that already has important implications for cave biogeography (see Magniez 1976).

The application of island biogeography theory to cave faunas is quite a different situation, however. The most reasonable position to take at this juncture is that there are insufficient data to either validate or falsify the hypothesis that caves or cave regions are virtual islands. There is no doubt that the cave environment is patchy on several geographic scales. What is in doubt is whether the fauna of caves or cave regions are in an equilibrium between immigration and extinction. On this point, more data are critically needed on turnover rates and species area curves.

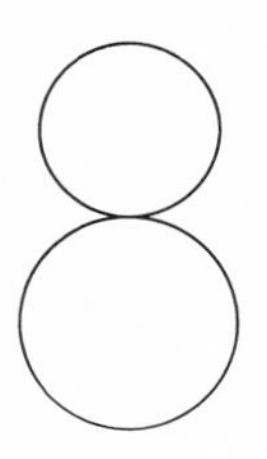

Future Directions

In several areas of evolutionary ecology the study of caves can play an increasingly important role. But to consider the possibilities, some mention must be made about the direction evolutionary ecology is likely to take in the next several decades. This question is receiving increasing attention from both evolutionary ecologists and philosophers of science, as evidenced by the recent symposium on "Conceptual Issues in Ecology" in the philosophical journal, *Synthese*. Although I have neither a well-worn copy of Popper's *The Logic of Scientific Discovery* nor a captive philosopher at hand, I will nonetheless make a brief foray into the battlefield, hoping to escape any direct hits from the heavy artillery present. Because of their different development, the fields of ecology and population genetics will be considered separately.

Ecology

A mature science of ecology will not have a thousand more or less unrelated models, each designed for a particular small group of organisms or habitats. On the other hand, models that have a formal, logical structure (usually mathematical) should play a central role. That is, there are general patterns and processes that are important for a large subset of species, communities, and habitats, although clearly we do not know what these patterns and processes are.

Over the past two decades the most popular ecological models have been of a kind that Levins (1966) describes as general and realistic but imprecise. Two prominent examples discussed in this book are community matrix theory (chapter 6), and island biogeography theory (chapter 7). The general research strategy of many ecologists has been to generate simple models, then test them in complex communities or search for concordances in complex communities. One gets the impression from much of the ecological literature that the ultimate test of a model occurs in the undisturbed tropics. At least in some prominent cases, this research strategy has failed (Simberloff 1980). There has been a failure to test data against the proper null hypothesis, and statistical reanalysis often indicates that the null hypothesis adequately explains the data. For examples in competition theory, see Strong (1980).

There is, it seems to me, an equally important question that has received less attention. Are complex systems, such as the tropics or organisms with complex life histories, the appropriate situations in which to test the models? In retrospect, the answer is no. This is not because the results when applied to complex systems have been disappointing, although they have been. But in complex situations, predictions of the model itself are qualitatively different from predictions in simple situations. For example, a two-species model of competition, using the much-maligned Lotka-Volterra equations of competition, predicts in the case of a stable equilibrium that the abundance of the competitors will be negatively correlated through time and that selection will tend to result in character displacement. But by adding even one more species using the same Lotka-Volterra competition equations, abundances can be either positively or negatively correlated, and either character convergence or displacement can occur, according to the theory. As a consequence, many tests of ecological models were misapplied because they used two-species models for multiple-species systems. Too often, these hypotheses were poorly tested or not really tested at all. But the demonstration that the hypothesis is falsified is not surprising, because not only the null hypothesis but the hypothesis itself was often ill formed.

A more appropriate place to test many ecological models is in relatively simple situations like caves. To reiterate, cave communities are simple, allowing more detailed examination of interactions; there are many replicate communities and natural occurrences of species additions and removals; and at least some assumptions of the models, such as near-equilibrium conditions, are more likely to be satisfied in caves than in more complex systems. But there have been weaknesses in the use of caves as model systems. The first is the virtual absence of experimental manipulations that would provide stronger tests of the theory. The second is the absence of long-term studies on fluctuations in age structure, population size, or species composition. These kinds of data are critical for the examination of underlying assumptions of equilibrium and steady state. The third weakness, which has been all too evident throughout this book, is the lack of sufficient data to make a strong test of a hypothesis. Data on area effect (chapter 7) and frequency of various species combinations (chapter 6) in caves will never be as extensive as data on, for example, birds on islands, because the species pool is much smaller. However the data base can be increased, and when it is it should be "cleaner," because in many cases transient species will be identifiable.

Suppose we take an optimistic look at the future and assume that the major ecological processes and patterns of cave faunas can be described and predicted by realistic, general models. Is there any reason to believe that this is also a step toward explaining more complex communities? We really don't know. Perhaps highly diverse communities such as those in the tropics have a completely different mode of organization, involving more mutualisms, higher-order interactions, and the like. But it seems equally, or even more, likely, that complex communities consist of relatively small subsets of strongly interacting species. Cave communities may then serve as a paradigm for a strongly interacting subset of species.

It is premature to conclude that the models of the last two decades have no validity. From the work reviewed in chapters 3, 6, and 7, on life history, species interactions, and island biogeography, some of these relatively simple models hold considerable promise. Cave biology can make contributions to the validation or falsification of these models in the following ways. First, long-term studies of population size and age structure could indicate the validity of some assumptions of the models and help sort out alternative explanations for the evolution of such phenomena as increased longevity and delayed reproduction. Second, perturbation experiments, especially species additions, could serve as a rigorous test of the species interaction models used in chapter 6. Third, monitoring of species turnover in caves could provide one of the best tests of island biogeography theory.

Population Genetics

The population genetics side of evolutionary ecology presents a different picture. Both the experimental and the theoretical work in population genetics is more sophisticated, but also less interrelated, than it is in ecology. For topics on complex genetic systems, such as linkage disequilibrium and the evolution of recombination, the number of theoretical papers greatly exceeds the number of experimental papers. Although experimental and theoretical population genetics need to be reconnected, it is highly unlikely that work on cave populations will play any role in most of this.

However, work on cave organisms may be important in reforging the connection between population genetics and adaptation. Since the use of gel electrophoresis to detect genetic variation, modern population genetics has become increasingly unconcerned with the phenotype above the level of enzymes. In the meantime, the study of morphological and physiological adaptation has, with some notable exceptions, fallen into decline because it no longer appears "modern." Both fields would benefit from a closer connection. Poulson's work on physiological and morphological adaptation in cave fish, Christiansen's on behavioral and morphological convergence in cave Collembola, and Wilkens' work on the genetics and morphology of regressive structures in cave fish provide an excellent starting place for considering the genetic basis of morphological and physiological change. A genetic analysis, using either the classical techniques of quantitative genetics or the modern techniques of gel electrophoresis, could begin to address problems of how much genetic change accompanies morphological and physiological changes within the context of well-defined morphological variation. It is an unsatisfactory state of affairs when adaptation and genetic variation are treated as wholly separate—as they are in this book and in the work of nearly all cave biologists.

Conservation

Anyone concerned about the protection and preservation of cave environments has doubtless cringed at some of my suggestions for species addition experiments and genetic analysis. Most cave biologists have been guilty of overcollecting at one time or another, and we all need to develop a greater sense of our effects on cave faunas. In my attempt to develop a stronger data base, I have caused a severe decline in the populations of several cave isopods. This problem is not easily resolved, but cave biologists should ask themselves if they *really* need the specimens they are about to collect, and whether any real advance in scientific knowledge will result from the collection.

References

Index

References

Absolon, K., and S. Hrabe. 1930. Über einen neuen Susswasser-Polychaeten aus den Höhlengewassern der Herzogowina. *Zoologischer Anzieger* 88:249–264.

Accordi, F., and V. Sbordoni. 1978. The fine structure of Hamann's organ in *Leptodirus hohenwarti*, a highly specialized cave Bathysciinae (Coleoptera, Catopidae). *International Journal of Speleology* 9:153–165.

Ambuhl, H. 1959. Die Bedeutung der Strömung als okologischer Faktor. *Schweiz Zeitschrift fur Hydrobiologie* 21:136–265.

Avise, J. C. 1976. Genetic differentiation during speciation. In *Molecular evolution*, ed. F. J. Ayala, pp. 106–122. Sunderland, Mass.: Sinauer Associates.

Avise, J. C., and R. K. Selander. 1972. Evolutionary genetics of cave-dwelling fishes of the genus *Astyanax*. *Evolution* 26:1–19.

Ayala, F. J. 1975. Genetic differentiation during the speciation process. *Evolutionary Biology* 8:1–78.

Barr, T. C. 1960. A synopsis of cave beetles of the genus *Pseudanophthalmus* of the Mitchell Plain in southern Indiana (Coleoptera, Carabidae). *American Midland Naturalist* 63:307–320.

——— 1965. The *Pseudanophthalmus* of the Appalachian Valley (Coleoptera: Carabidae). *American Midland Naturalist* 73:41–72.

——— 1967a. Ecological studies in the Mammoth Cave system of Kentucky. I. The biota. *International Journal of Speleology* 3:147–203.

——— 1967b. A new *Pseudanophthalmus* from an epigean environment in West Virginia (Coleoptera: Carabidae). *Psyche* 74:166–172.

——— 1967c. Observations on the ecology of caves. *American Naturalist* 101:475–492.

——— 1968. Cave ecology and the evolution of troglobites. *Evolutionary Biology* 2:35–102.

——— 1969. Evolution of the Carabidae (Coleoptera) in the southern Appalachians. In *The distributional history of the biota of the southern Appalachians. Pt. I. Invertebrates,* ed. P. C. Holt, pp. 67–92. Blacksburg, Va.: Virginia Polytechnic Institute Press.

——— 1974a. Revision of *Rhadine* LeConte (Coleoptera, Carabidae) I. The *subterranea* group. *American Museum Novitates* no. 2539.

——— 1974b. The eyeless beetles of the genus *Arianops* Brendel (Coleoptera, Pselaphidae). *American Museum of Natural History Bulletin* 154:3–51.

——— 1979. The taxonomy, distribution, and affinities of *Neaphaenops,* with notes on associated species of *Pseudanophthalmus* (Coleoptera, Carabidae). *American Museum Novitates* no. 2682.

——— 1981. *Pseudanophthalmus* from Appalachian caves (Coleoptera: Carabidae): the *engelhardti* complex. *Brimleyana* no. 5:37–94.

Barr, T. C., and P. H. Crowley. 1981. Do cave carabid beetles really show character displacement in body size? *American Naturalist* 117:363–371.

Barr, T. C., and C. H. Krekeler. 1967. *Xenotrechus,* a new genus of cave trechines from Missouri (Coleoptera: Carabidae). *Annals of the Entomological Society of America* 60:1322–1325.

Barr, T. C., and R. A. Kuehne. 1971. Ecological studies in the Mammoth Cave ecosystems of Kentucky. II. The ecosystem. *Annales de Spéléologie* 26:47–96.

Barr, T. C., and S. B. Peck. 1965. Occurrence of a troglobitic *Pseudanophthalmus* outside a cave (Coleoptera: Carabidae). *American Midland Naturalist* 73:73–74.

Beatty, R. A. 1949. The pigmentation of cavernicolous animals. III. The carotenoid pigments of some amphipod crustacea. *Journal of Experimental Biology* 26:125–130.

Besharse, J. C., and R. A. Brandon. 1973. Optomotor response and eye structure of the troglobitic salamander *Gyrinophilus palleucus. American Midland Naturalist* 89:463–467.

Besharse, J. C., and J. R. Holsinger. 1977. *Gyrinophilus subterraneus,* a new troglobitic salamander from southern West Virginia. *Copeia* (1977):624–634.

Bögli, A. 1980. *Karst hydrology and physical speleology,* trans. J. C. Schmid. Berlin: Springer-Verlag.

Bou, C., and R. Rouch. 1967. Un nouveau champ de recherches sur la faune aquatique souterraine. *Compte Rendus Académie des Sciences* 265D: 369–370.

Bousfield, E. L. 1958. Fresh water amphipod crustaceans of glaciated North America. *Canadian Field Naturalist* 72:55–113.

——— 1977. A new look at the systematics of gammaroidean amphipods of the world. *Crustaceana* (supplement) 4:282–316.

Brandon, R. A. 1971. North American troglobitic salamanders: some apects of modification in cave habitats, with special reference to *Gyrinophilus palleucus. National Speleological Society Bulletin* 33:1–22.

Breder, C. M., Jr. 1942. Descriptive ecology of La Cueva Chica, with especial reference to the blind fish, *Anoptichthys*. *Zoologica* 27:7–16.

——— 1944. Ocular anatomy and light sensitivity studies on the blind fish from Cueva de los Sabinos, Mexico. *Zoologica* 29:131–143.

——— 1953. Cave fish evolution. *Evolution* 7:179–181.

Brown, J. 1971. Mammals on mountaintops: nonequilibrium insular biogeography. *American Naturalist* 105:467–478.

Bruce, R. C. 1979. Evolution of paedomorphosis in salamanders of the genus *Gyrinophilus*. *Evolution* 33:998–1000.

Brunner, G., and T. C. Kane. 1981. The ecological genetics of four subspecies of *Neaphaenops tellkampfi* (Coleoptera: Carabidae). *Proceedings of the Eighth International Congress of Speleology, Bowling Green, Kentucky* 1:48–49.

Burbanck, W. D., J. P. Edwards, and M. P. Burbanck. 1948. Toleration of lowered oxygen tension by cave and stream crayfish. *Ecology* 29:360–367.

Bush, G. L. 1975. Modes of animal speciation. *Annual Review of Ecology and Systematics* 6:339–364.

Caswell, H., and P. A. Werner. 1978. Transient behavior and life history analysis of teasel (*Dipsacus sylvestris* Hud.). *Ecology* 59:53–66.

Caumartin, V. 1963. Review of the microbiology of underground environments. *National Speleological Society Bulletin* 25:1–14.

Charlesworth, B. 1980. *Evolution in age-structured populations*. Cambridge: Cambridge University Press.

Charlesworth, B., and D. Charlesworth. 1973. A study of linkage disequilibrium in populations of *Drosophila melanogaster*. *Genetics* 73:351–359.

Charnov, E. L., and W. M. Schaffer. 1973. Life history consequences of natural selection: Cole's result revisited. *American Naturalist* 107:791–793.

Christiansen, K. A. 1961. Convergence and parallelism in cave Entomobryinae. *Evolution* 15:288–301.

——— 1965. Behavior and form in the evolution of cave Collembola. *Evolution* 19:529–537.

Christiansen, K. A., and P. Bellinger. 1980–81. *The Collembola of North America north of the Rio Grande*. Grinnell, Iowa: Grinnell College.

Christiansen, K. A., and M. Bullion. 1978. An evolutionary and ecological analysis of the terrestrial arthropods of caves in the central Pyrenees. I. Ecological analysis with special reference to Collembola. *National Speleological Society Bulletin* 40:103–117.

Christiansen, K. A., and D. C. Culver. 1968. Geographical variation and evolution in *Pseudosinella hirsuta*. *Evolution* 22:237–255.

Cockley, D. E., J. L. Gooch, and D. P. Weston. 1977. Genic diversity in cave-dwelling crickets (*Ceuthophilus gracilipes*). *Evolution* 31:313–318.

Connell, J. H. 1975. Some mechanisms producing structure in natural communities: a model and evidence from field experiments. In *Ecology and evolution of communities*, ed. M. L. Cody and J. M. Diamond, pp. 460–490. Cambridge, Mass.: Harvard University Press.

Connor, E. F., and E. D. McCoy. 1979. The statistics and biology of the species-area relationship. *American Naturalist* 113:791–833.

Cooper, J. E. 1975. Ecological and behavioral studies in Shelta Cave, Alabama, with emphasis on decapod crustaceans. Ph.D. dissertation, University of Kentucky.

Cooper, J. E., and R. A. Kuehne. 1974. *Speoplatyrhinus poulsoni,* a new genus and species of subterranean fish from Alabama. *Copeia* (1974):486–493.

Cooper, M. R. 1969. Sensory specialization and allometric growth in cavernicolous crayfishes. *Proceedings of the Fourth International Congress of Speleology, Ljubljana, Yugoslavia* 4–5:203–208.

Coyne, J. A. 1976. Lack of genic similarity between two sibling species of *Drosophila* as revealed by varied techniques. *Genetics* 84:593–607.

Coyne, J. A., W. Eanes, and R. C. Lewontin. 1979. The genetics of electrophoretic variation. *Genetics* 92:353–361.

Crawford, R. L. 1981. A critique of the analogy between caves and islands. *Proceedings of the Eighth International Congress of Speleology, Bowling Green, Kentucky* 1:295–297.

Culver, D. C. 1970a. Analysis of simple cave communities: niche separation and species packing. *Ecology* 51:949–958.

——— 1970b. Analysis of simple cave communities. I. Caves as islands. *Evolution* 24:463–474.

——— 1971a. Analysis of simple cave communities. III. Control of abundance. *American Midland Naturalist* 85:173–187.

——— 1971b. Caves as archipelagoes. *National Speleological Society Bulletin* 33:97–100.

——— 1973a. Competition in spatially heterogeneous systems: an analysis of simple cave communities. *Ecology* 54:102–110.

——— 1973b. Feeding behavior of the salamander *Gyrinophilus porphyriticus* in caves. *International Journal of Speleology* 5:369–377.

——— 1974. Competition between Collembola in a patchy environment. *Revue de Écologie et Biologie du Sol* 11:533–540.

——— 1975. The interaction of competition and predation in cave stream communities. *International Journal of Speleology* 7:229–245.

——— 1976. The evolution of aquatic cave communities. *American Naturalist* 110:945–957.

——— 1978. Cave communities and statistical inference: a reply. *American Naturalist* 112:160–161.

——— 1981. Some implications of competition for cave stream communities. *International Journal of Speleology* 11:49–62.

Culver, D. C., and T. J. Ehlinger. 1980. The effects of microhabitat size and competitor size on two cave isopods. *Brimleyana* no. 4:103–114.

Culver, D. C., J. R. Holsinger, and R. A. Baroody. 1973. Toward a predictive cave biogeography: the Greenbrier Valley as a case study. *Evolution* 27:689–695.

Culver, D. C., and T. L. Poulson. 1971. Oxygen consumption and activity in closely related amphipod populations from cave and surface habitats. *American Midland Naturalist* 85:74–84.

Curl, R. 1966. Caves as a measure of karst. *Journal of Geology* 74:798–830.

Danielopol, D. L. 1980. An essay to assess the age of the freshwater interstitial ostracods of Europe. *Bijdrugen tot de Dierkunde* 50:243–291.

Darwin, C. 1859. *On the origin of species by means of natural selection, or the preservation of favoured races in the struggle of life*. London: John Murray.

Davidson, D. W. 1980. Some consequences of diffuse competition in a desert ant community. *American Naturalist* 116:92–105.

Dayton, P. K. 1973. Two cases of resource partitioning in an intertidal community: making the right prediction for the wrong reason. *American Naturalist* 107:662–670.

Deeleman-Reinhold, C. L. 1981. Remarks on the origin and distribution of troglobitic spiders. *Proceedings of the Eighth International Congress of Speleology, Bowling Green, Kentucky* 1:305–308.

Delay, B., V. Sbordoni, M. Cobolli-Sbordoni, and E. de Matthaeis. 1980. Divergences génétiques entre les populations de *Speonomus delarouzeei* du Massif du Canigou (Coleoptera, Bathysciinae). *Mémoires de Biospeologie* 7:235–247.

Deleurance-Glaçon, S. 1963. Recherches sur les coléoptères troglobies de la sous-famille des Bathysciinae. *Annales des Sciences Naturelles, Zoologie, Paris. 12e serie*. 5:1–172.

Derouet, L. 1959. Contribution à l'étude de la biologie de deux crustaces aquatiques cavernicoles: *Caecosphaeroma burgundum* et *Niphargus orcinus virei*. *Vie et Milieu* 10:321–346.

Diamond, J. M. 1975. Assembly of species communities. In *Ecology and evolution of communities*, ed. M. L. Cody and J. M. Diamond, pp. 342–444. Cambridge, Mass.: Harvard University Press.

Dickson, G. W. 1976. Variation in the ecology, morphology and behavior of the troglobitic amphipod crustacean *Crangonyx antennatus* Packard (Crangonychidae) from different habitats. M.S. thesis, Old Dominion University, Norfolk, Va.

——— 1977. Variation among populations of the troglobitic amphipod crustacean *Crangonyx antennatus* Packard living in different habitats. I. Morphology. *International Journal of Speleology* 9:43–58.

Dickson, G. W., and R. Franz. 1980. Respiration rates, ATP turnover and adenylate energy charge in excised gills of surface and cave crayfish. *Comparative and Biochemical Physiology* 65A:375–379.

Dickson, G. W., and P. W. Kirk. 1976. Distribution of heterotrophic microorganisms in relation to detritivores in Virginia caves (with supplemental bibliography on cave mycology and microbiology). In *The distributional history of the biota of the southern Appalachians. IV. Algae and fungi*. ed. B. C. Parker and M. K. Roane, pp. 205–226. Charlottesville, Va.: University of Virginia Press.

Dickson, G. W., J. C. Patton, J. R. Holsinger, and J. C. Avise. 1979. Genetic variation in cave-dwelling and deep-sea organisms, with emphasis on *Crangonyx antennatus* (Crustacea: Amphipoda) in Virginia. *Brimleyana* no. 2:119–130.

Dykhuizen, D. 1978. Selection for tryptophan autotrophs of *Escherischia coli* in glucose-limited chemostats as a test of the energy conservtion hypothesis of evolution. *Evolution* 32:125–150.

Eigenmann, C. H. 1909. Cave vertebrates of America, a study in degenerative evolution. *Carnegie Institute of Washington Publications* no. 104.

Endler, J. A. 1977. *Geographic variation, speciation, and clines*. Princeton: Princeton University Press.

Estes, J. A. 1978. The comparative ecology of two populations of the troglobitic isopod crustacean *Lirceus usdagalun* (Asellidae). M.S. thesis, Old Dominion University, Norfolk, Va.

Ewens, W. J., and M. W. Feldman. 1976. The theoretical assessment of selective neutrality. In *Population genetics and ecology*, ed. S. Karlin and E. Nevo, pp. 303–339. New York: Academic Press.

Falconer, D. S. 1960. *Introduction to quantitative genetics*. New York: Ronald Press.

Finnerty, V., and G. Johnson. 1979. Post-translational modification as a potential explanation of high levels of enzyme polymorphism: xanthine dehydrogenase and aldehyde oxidase in *Drosophila melanogaster*. *Genetics* 91:695–722.

Franz, R., and D. S. Lee. In press. Distribution and evolution of Florida's troglobitic crayfishes. *International Journal of Speleology*.

Gantmacher, F. R. 1959. *Applications of the theory of matrices*. New York: Wiley Interscience.

Gilbert, J., and J. Mathieu. 1980. Relations entre les teneurs en proteines, glucides et lipides au cours du jeûne expérimental, chez deux espèces de *Niphargus* peuplant des biotopes differents. *Crustaceana* (supplement) 6:137–147.

Gilbert, F. S. 1980. The equilibrium theory of island biogeography: fact or fiction? *Journal of Biogeography* 7:209–235.

Ginet, R. 1960. Écologie, éthologie et biologie de *Niphargus* (Amphipodes Gammaridés hypogés). *Annales de Spéléologie* 15:1–254.

Gittleson, S. M., and R. L. Hoover. 1970. Protozoa of underground waters in caves. *Annales de Spéléologie* 25:91–106.

Giuseffi, S., T. C. Kane, and W. F. Duggleby. 1978. Genetic variability in the Kentucky cave beetle *Neaphaenops tellkampfi* (Coleoptera: Carabidae). *Evolution* 32:679–681.

Gooch, J. L., and S. W. Golladay. 1981. Genetic population structure in an amphipod species. *International Journal of Speleology* 11:15–20.

Gooch, J. L., and S. W. Hetrick. 1979. The relation of genetic structure to environmental structure: *Gammarus minus* in a karst area. *Evolution* 33:192–206.

Goodman, D. 1974. Natural selection and a cost ceiling on reproductive effort. *American Naturalist* 108:247–268.

Gould, S. J. 1971. Geometric similarity in allometric growth: a contribution to the problem of scaling in the evolution of size. *American Naturalist* 105:113–136.

——— 1977. *Ontogeny and phylogeny*. Cambridge, Mass.: Harvard University Press.

Gould, S. J., and R. C. Lewontin. 1979. The spandrels of San Marcos and the Panglossian paradigm: a critique of the adaptationist programme. *Proceedings of the Royal Society of London, Series B* 205:581–598.

Gresser, E. B., and C. M. Breder, Jr. 1940. The histology of the eye of the cave characin, *Anoptichthys*. *Zoologica* 25:113–116.

Halliday, W. R. 1974. *American caves and caving: techniques, pleasures, and safeguards of modern cave exploration*. New York: Harper and Row.

Hamilton-Smith, E. 1971. The classification of cavernicoles. *National Speleological Society Bulletin* 33:63–66.

Harris, H. 1966. Enzyme polymorphism in man. *Proceedings of the Royal Society of London, Series B* 164:298–310.

Henry, J. P. 1978. Observations sur les peuplements de Crustacés Aselloides des milieux souterrains. *Bulletin de la Société Zoologique de France* 103:491–497.

Herbers, J. M. 1980. The evolution of sex-ratio strategies in hymenopteran societies. *American Naturalist* 114:818–834.

Herwig, H. J. 1976. Comparative ultrastructural investigations of the pineal organ of the blind cave fish, *Anoptichthys jordani*, and its ancestor, the eyed river fish, *Astyanax mexicanus*. *Cell and Tissue Research* 167:297–324.

Hobbs, H. H. II, and T. C. Barr. 1972. Origins and affinities of the troglobitic crayfishes of North America (Decapoda: Astacidae). II. Genus *Orconectes*. *Smithsonian Contributions to Zoology* no. 105.

Hobbs, H. H. III. 1973. The population dynamics of cave crayfishes and their commensal ostracods from southern Indiana. Ph.D. thesis, Indiana University.

——— 1975. Distribution of Indiana cavernicolous crayfishes and their ectocommensal ostracods. *International Journal of Speleology* 7:273–302.

Holsinger, J. R. 1966. A preliminary study of the effects of organic pollution of Banners Corner Cave, Virginia. *International Journal of Speleology* 2:75–89.

——— 1969. Biogeography of the freshwater amphipod crustaceans (Gammaridae) of the central and southern Appalachians. In *The distributional history of the southern Appalachians. I; Invertebrates*, ed. P. C. Holt, pp. 19–50. Blacksburg, Va.: Virginia Polytechnic Institute Press.

——— 1971. A new species of the subterranean amphipod genus *Allocrangonyx* (Gammaridae), with a redescription of the genus and remarks on its zoogeography. *International Journal of Speleology* 3:317–331.

——— 1972. *The freshwater amphipod crustaceans (Gammaridae) of North America*. Biota of Freshwater Ecosystems Identification Manual no. 5. Washington, D. C.: Environmental Protection Agency.

——— 1974. Systematics of the subterranean amphipod genus *Stygobromus* (Gammaridae), pt. I. Species of the western United States. *Smithsonian Contributions to Zoology* no. 160.

——— 1975. *Descriptions of Virginia caves*. Virginia Division of Mineral Resources, Bulletin 85. Charlottesville, Va.

——— 1978. Systematics of the subterranean amphipod genus *Stygobromus* (Crangonyctidae). II. Species of the eastern United States. *Smithsonian Contributions to Zoology* no. 266.

——— 1981. *Stygobromus canadensis*, a troglobitic amphipod crustacean from Castleguard Cave, with remarks on the concept of cave glacial refugia. *Proceedings of the Eighth International Congress of Speleology, Bowling Green, Kentucky* 1:93–95.

Holsinger, J. R., R. A. Baroody, and D. C. Culver. 1975. The invertebrate cave fauna of West Virginia. *West Virginia Speleological Survey Bulletin* no. 7.

Holsinger, J. R., and T. E. Bowman. 1973. A new troglobitic isopod of the genus *Lirceus* (Asellidae) from southwestern Virginia, with notes on its ecology and additional cave records for the genus in the Appalachians. *International Journal of Speleology* 5:261–271.

Holsinger, J. R., and D. C. Culver. 1970. Morphological variation in *Gammarus minus* Say (Amphipoda, Gammaridae) with emphasis on subterranean forms. *Postilla* no. 146.

Holsinger, J. R., and G. W. Dickson. 1977. Burrowing as a means of survival in the troglobitic amphipod crustacean *Crangonyx antennatus* Packard (Crangoyctidae). *Hydrobiologia* 54:195–199.

Holsinger, J. R., and G. Longley. 1980. The subterranean amphipod fauna of an artesian well in Texas. *Smithsonian Contributions to Zoology* no. 308.

Howarth, F. G. 1972. Cavernicoles in lava tubes on the island of Hawaii. *Science* 175:325–326.

——— 1980. The zoogeography of specialized cave animals: a bioclimatic model. *Evolution* 34:394–406.

Hubbell, T. H., and R. M. Norton. 1978. The systematics and biology of the cave-crickets of the North American tribe Hadenicini (Orthoptera: Saltatoria: Ensifera: Rhaphidophoridae: Dolichopodinae). *Miscellaneous Publications of the Museum of Zoology, University of Michigan* no. 156.

Hubricht, L. 1976. The genus *Fontigens* from Appalachian caves. *The Nautilus* 90:86–88.

Jegla, T. C., and T. L. Poulson. 1968. Evidence of circadian rhythms in a cave crayfish. *Journal of Experimental Zoology* 168:273–282.

Jennings, J. N. 1971. *Karst*. Cambridge, Mass.: M.I.T. Press.

Juberthie, C. 1969. Relations entre le climat, le microclimat et les *Aphaenops cerberus* dans la grotte de Sainte-Catherine (Ariege). *Annales de Spéléologie* 24:75–104.

Juberthie, C., and B. Delay. 1981. Ecological and biological implications of a "superficial underground compartment." *Proceedings of the Eighth International Congress of Speleology, Bowling Green, Kentucky* 1:203–205.

Juberthie, C., B. Delay, and M. Bouillon. 1980. Extension du milieu souterrain en zone non-calcaire: description d'un nouveau milieu et de son peuplement par les Coléoptères troglobies. *Mémoires de Biospeologie* 7:19–52.

Kane, T. C. 1974. Studies of simple cave communities: predation strategies of two co-occurring carabid beetles. Ph.D. dissertation, University of Notre Dame.

——— In press. Genetic patterns and population structure in cave animals. In *Environmental Adaptation and Evolution in Arthropods and Lower Vertebrates*. Bremen Symposium on Biological Systems Theory.

Kane, T. C., R. M. Norton, and T. L. Poulson. 1975. The ecology of a predaceous troglobitic beetle, *Neaphaenops tellkampfii* (Coleoptera: Carabidae, Trechinae). I. Seasonality of food input and early life history stages. *International Journal of Speleology* 7:45–54.

Kane, T. C., and T. L. Poulson. 1976. Foraging by cave beetles: spatial and temporal heterogeneity of prey. *Ecology* 57:793–800.

Keith, J. H. 1975. Seasonal changes in a population of *Pseudanophthalmus tenuis* (Coleoptera, Carabidae) in Murray Spring Cave, Indiana: a preliminary report. *International Journal of Speleology* 7:33–44.

Kimura, M., and T. Ohta. 1971. *Theoretical aspects of population genetics.* Princeton: Princeton University Press.

King, R. C. 1974. *A dictionary of genetics.* New York: Oxford University Press.

Kosswig, C. 1948. Genetische Beiträge zur Praadaptations-theorie. *Revue de Facultie des Science (Istanbul) Series B* 5:176–209.

——— 1965. Génétique et évolution régressive. *Revue des Questions Scientifiques* 136:227–257.

Kosswig, C., and L. Kosswig. 1940. Die variabilität bei *Asellus aquaticus* unter besonderer Berucksichtigung der Variabilität in isolierten unter-und aberirdischen Population. *Revue de Facultie des Science (Istanbul) Series B* 5:1–55.

Laing, C. D., G. R. Carmody, and S. B. Peck. 1976a. How common are sibling species in cave-inhabiting invertebrates? *American Naturalist* 110: 184–189.

——— 1976b. Population genetics and evolutionary biology of the cave beetle *Ptomaphagus hirtus. Evolution* 30:484–489.

Lande, R. 1976. Natural selection and random genetic drift in phenotypic evolution. *Evolution* 30:314–334.

——— 1978. Evolutionary mechanisms of limb loss in tetrapods. *Evolution* 32:73–92.

Laneyrie, R. 1967. Nouvelle classification des Bathysciinae (Coléoptères Catopidae). *Annales des Spéléologie* 22:587–645.

Lawlor, L. R. 1980. Structure and stability in natural and randomly constructed competitive communities. *American Naturalist* 116:394–408.

Leigh Brown, A. J., and C. H. Langley. 1979. Reevaluation of level of genic heterozygosity in natural populations of *Drosophila melanogaster* by 2-dimensional electrophoresis. *Proceedings of the National Academy of Science (U.S.A.)* 76:2381–2384.

Leopold, L. B., M. G. Wolman, and J. P. Miller. 1964. *Fluvial processes in geomorphology.* San Francisco: W. H. Freeman.

Levin, S. A. 1974. Dispersion and population interactions. *American Naturalist* 108:207–228.

Levine, S. H. 1976. Competitive interactions in ecosystems. *American Naturalist* 110:903–910.

Levins, R. 1966. The strategy of model building in population biology. *American Scientist* 54:421–431.

——— 1975. Evolution in communities near equilibrium. In *Ecology and evolution of communities,* ed. M. L. Cody and J. M. Diamond, pp. 16–50. Cambridge, Mass.: Harvard University Press.

Lewontin, R. C. 1974. *The genetic basis of evolutionary change.* New York: Columbia University Press.

——— 1979. Fitness, survival, and optimality. In *Analysis of ecological systems,* ed. D. J. Horn, G. R. Stairs, and R. D. Mitchell, pp. 3–22. Columbus: Ohio State University Press.

Lewontin, R. C., and J. L. Hubby. 1966. A molecular approach to the study of genic heterozygosity in natural populations. II. Amount of variation and degree of heterozygosity in natural populations of *Drosophila pseudoobscura*. *Genetics* 54:595–609.

Lin, E. C. C., A. J. Hacking, and J. Aguilar. 1976. Experimental models of acquisitive evolution. *Bioscience* 26:548–555.

Lucarelli, M., and V. Sbordoni. 1978. Humidity responses and the role of Hamann's organ of cavernicolous Bathysciinae. *International Journal of Speleology* 9:167–177.

MacArthur, R. H. 1962. Some generalized theorems of natural selection. *Proceedings of the National Academy of Science (U.S.A.)* 38:1893–1897.

MacArthur, R. H., and R. Levins. 1967. The limiting similarity, convergence, and divergence of coexisting species. *American Naturalist* 101:377–385.

MacArthur, R. H., and E. O. Wilson. 1967. *The theory of island biogeography*. Princeton: Princeton University Press.

MacDonald, N. 1978. *Time lags in biological models*. Lecture Notes in Biomathematics, vol. 27. New York: Springer-Verlag.

Magniez, G. 1976. Remarques sur la biologie et l'ecologie de *Stenasellus virei* Dollfus (Crustacea Isopoda Asellota des eaux souterraines). *International Journal of Speleology* 8:135–140.

Maguire, B. 1961. Regressive evolution in cave animals and its mechanism. *Texas Journal of Science* 13:363–370.

Mankin, J. B., R. V. O'Neill, H. H. Shugart, and B. W. Rust. 1977. The importance of validation in ecosystem analysis. In *New directions in the analysis of ecological systems, I*. ed. G. S. Innis, pp. 63–72. La Jolla, Calif.: Society for Computer Simulation.

Maruyama, T., and M. Kimura. 1978. Theoretical study of genetic variability, assuming step-wise production of neutral and very slightly deleterious mutations. *Proceedings of the National Academy of Science (U.S.A.)* 75:919–922.

Mateu, J. 1980. Commentaires sur deux Agonini troglobies de l'Amérique centrale et méridionale. *Mémoires de Biospeologie* 7:209–213.

Mathieu, J. 1980. Activité locomotrice et métabolisme respiratore à 11°C de l'Amphipode troglobie *Niphargus rhenorhodanensis* Schellenberg, 1937. *Crustaceana* (supplement) 6:160–169.

Matjašič, J. 1958. Biologie und zoogeographie der europäischen Temnocephaliden. *Zooligischer Anzieger* (supplementband) 21:477–482.

May, R. M. 1972. Limit cycles in predator-prey communities. *Science* 177:900–902.

——— 1975. Patterns of species abundance and diversity. In *Ecology and evolution of communities,* ed. M. L. Cody and J. M. Diamond, pp. 81–120. Cambridge, Mass.: Harvard University Press.

Maynard Smith, J. 1978. Optimization theory in evolution. *Annual Review of Ecology and Systematics* 9:31–56.

McKinney, T. 1975. Studies on the niche separation in two carabid cave beetles. *International Journal of Speleology* 7:65–78.

Mertz, D. B. 1971. Life history phenomena in increasing and decreasing populations. In *Statistical ecology, vol. 2,* ed. G. P. Patil, E. C. Pielou, and

W. E. Waters, pp. 361–400. University Park, Pa.: Pennsylvania State University Press.

Mitchell, R. W. 1968. Food and feeding habits of the troglobitic carabid beetle *Rhadine subterranea*. *International Journal of Speleology* 3:249–270.

——— 1969. A comparison of temperate and tropical cave communities. *Southwestern Naturalist* 14:73–88.

——— 1970. Total number and density estimates of some species of cavernicoles inhabiting Fern Cave, Texas. *Annales de Spéléologie* 25:73–90.

Mitchell, R. W., W. H. Russell, and W. R. Elliott. 1977. Mexican eyeless characin fishes, genus *Astyanax:* environment, distribution, and evolution. *Special Publications. The Museum, Texas Tech University* no. 12.

Muchmore, W. B. 1976. New cavernicolous species of *Kleptochthonius,* and recognition of a new species group within the genus (Pseudoscorpionida: Chthoniidae). *Entomological News* 87:211–217.

Nei, M. 1972. Genetic distance between populations. *American Naturalist* 106:283–292.

——— 1975. *Molecular population genetics and evolution*. New York: American Elsevier.

Neill, W. E. 1974. The community matrix and interdependence of competition coefficients. *American Naturalist* 108:399–408.

Nelson, G., and N. Platnick. 1981. *Systematics and biogeography. Cladistics and vicariance*. New York: Columbia University Press.

Nevo, E. 1976. Adaptive strategies of genetic systems in constant and varying environments. In *Populations genetics and ecology,* ed. S. Karlin and E. Nevo, pp. 141–158. New York: Academic Press.

Norton, R. M., T. C. Kane, and T. L. Poulson. 1975. The ecology of a predaceous troglobitic beetle, *Neaphaenops tellkampfii* (Coleoptera: Carabidae, Trechinae). II. Adult seasonality, feeding and recruitment. *International Journal of Speleology* 7:55–64.

Packard, A. S. 1888. The cave fauna of North America with remarks on the anatomy of the brain and origin of the blind species. *Memoirs of the National Academy of Sciences,* vol. 4, pt I. Reprint, New York: Arno Press, 1977.

——— 1901. *Lamarck, the founder of evolution*. New York: Longmans, Green.

Paine, R. T., and S. A. Levin. 1981. Intertidal landscapes: disturbance and the dynamics of pattern. *Ecological Monographs* 51:145–178.

Park, O. 1951. Cavernicolous pselaphid beetles of Alabama and Tennessee, with observations on the taxonomy of the family. *Geological Survey of Alabama, Museum Paper* no. 31.

Peck, S. B. 1971. The invertebrate fauna of tropical American caves. I. Chilibrillo Cave, Panama. *Annales de Spéléologie* 26:423–437.

——— 1973a. A systematic revision and the evolutionary biology of the *Ptomaphagus* (*Adelops*) beetles of North America (Coleoptera; Leiodidae; Catopinae), with emphasis on cave-inhabiting species. *Museum of Comparative Zoology Bulletin* 145:29–162.

——— 1973b. Feeding efficiency in the cave salamander *Haideotriton wallacei*. *International Journal of Speleology* 5:15–19.

——— 1975a. The life cycle of a Kentucky cave beetle, *Ptomaphagus hirtus* (Coleoptera; Leiodidae; Catopinae). *International Journal of Speleology* 7:7–17.

——— 1975b. A population study of the cave beetle *Ptomaphagus loedingi* (Coleoptera; Leiodidae; Catopinae). *International Journal of Speleology* 7:19–32.

——— 1977. An unusual sense receptor in internal vesicles of *Ptomaphagus* (Coleoptera: Leiodidae). *Canadian Entomologist* 109:81–86.

——— 1980. Climatic change and the evolution of cave invertebrates in the Grand Canyon, Arizona. *National Speleological Society Bulletin* 42:53–60.

——— 1981a. The geological, geographical, and environmental setting of cave faunal evolution. *Proceedings of the Eighth International Congress of Speleology, Bowling Green, Kentucky* 2:501–502.

——— 1981b. Evolution of cave Cholevinae in North America (Coleoptera: Leiodidae). *Proceedings of the Eighth International Congress of Speleology, Bowling Green, Kentucky* 2:503–505.

Peck, S. B., and B. L. Richardson. 1976. Feeding ecology of the salamander *Eurycea lucifuga* in the entrance, twilight, and dark zones of caves. *Annales de Spéléologie* 31:175–182.

Peck, S. B., and D. R. Russell. 1976. Life history of the fungus gnat *Macrocera nobilis* in American caves (Diptera: Mycetophilidae). *Canadian Entomologist* 108:1235–1241.

Peckham, M., ed. 1959. *The Origin of Species by Charles Darwin: a variorum text*. Philadelphia: University of Pennsylvania Press.

Pimm, S. L. 1978. Cave communities and statistical inference. *American Naturalist* 112:159–160.

Poulson, T. L. 1963. Cave adaptation in amblyopsid fishes. *American Midland Naturalist* 70:257–290.

——— 1964. Animals in aquatic environments: animals in caves. In *Handbook of physiology*, ed. D. B. Dill, pp. 749–771. Washington, D.C.: American Physiological Society.

——— 1969. Population size, density, and regulation in cave fishes. *Proceedings of the Fourth International Congress of Speleology, Ljubljana, Yugoslavia* 4–5:189–192.

——— 1978. Community organization. In *Cave Research Foundation 1978 Annual Report*, ed. S. G. Wells, pp. 41–45. Yellow Springs, Ohio.

——— 1981. Variations in life history of linyphiid cave spiders. *Proceedings of the Eighth International Congress of Speleology, Bowling Green, Kentucky* 1:60–62.

Poulson, T. L., and W. B. White. 1969. The cave environment. *Science* 165:971–981.

Price, P. W. 1980. *Evolutionary biology of parasites*. Princeton: Princeton University Press.

Prout, T. 1964. Observations on structural reduction in evolution. *American Naturalist* 97:239–249.

Reddell, J. R. 1981. A review of the cavernicole fauna of Mexico, Guatemala, and Belize. *Texas Memorial Museum Bulletin* 27:1–327.

Reynoldson, T. B., and L. S. Bellamy. 1971. The establishment of interspecific competition in field populations, with an example of competition in action between *Polycelis nigra* (Mull.) and *P. tenuis* (Ijima) (Turbellaria, Tricladida). In *Dynamics of Populations, Proceedings of the Advanced Institute on Dynamics of Numbers in Populations,* ed. P. J. Dan Boer and G. R. Gradwell, pp. 282–297. Wageningen, Netherlands: Pudoc.

Risch, S., and D. H. Boucher. 1976. What ecologists look for. *Ecological Society of America Bulletin* 57:8–9.

Rosenzweig, M. L., and R. H. MacArthur. 1963. Graphical representation and stability conditions of predator-prey interactions. *American Naturalist* 97:209–223.

Rouch, R. 1968. Contribution à la connaissance des harpacticides hypogés (Crustacés-Copepodes). *Annales de Spéléologie* 23:5–167.

Roughgarden, J. 1979. *Theory of population genetics and evolutionary ecology: an introduction.* New York: Macmillan.

Şadoglu, P. 1957. Mendelian inheritance in the hybrids between the Mexican blind cave fishes and their overground ancestor. *Verhandlungen Deutsche Zoologische Gesellschaft, Graz* (1957):432–439.

——— 1967. The selective value of eye and pigment loss in Mexican cave fish. *Evolution* 21:541–549.

Sbordoni, V., G. Allegrucci, A. Caccone, D. Caesaroni, M. Cobolli-Sbordoni, and E. de Matthaeis. 1981. Genetic variability and divergence in cave populations of *Troglophilus cavicola* and *T. andreinii* (Orthoptera, Rhaphidophoridae). *Evolution* 35:226–233.

Sbordoni, V., A. Caccone, E. de Matthaeis, and M. Cobolli-Sbordoni. 1980. Biochemical divergence between cavernicolous and marine Sphaeromidae and the Mediterranean salinity crisis. *Experientia* 36:48–50.

Schaffer, W. M. 1974a. Optimal reproductive effort in fluctuating environments. *American Naturalist* 108:783–790.

——— 1974b. Selection for optimal life histories: the effects of age structure. *Ecology* 55:291–303.

Schlagel, S. R., and C. M. Breder, Jr. 1947. A study of the oxygen consumption of blind and eyed cave characins in light and in darkness. *Zoologica* 32:17–28.

Schultz, G. A. 1970. Descriptions of new subspecies of *Ligidium elrodii* (Packard) comb. nov. with notes on other isopod crustaceans from caves in North America (Oniscoidea). *American Midland Naturalist* 84:36–45.

Selander, R. K. 1976. Genic variation in natural populations. In *Molecular evolution,* ed. J. Ayala, pp. 21–45. Sunderland, Mass.: Sinauer Associates.

Shear, W. A. 1972. Studies on the milliped order Chordeumida (Diplopoda): a revision of the family Cleidogonidae and a reclassification of the order Chordeumida in the New World. *Museum of Comparative Zoology Bulletin* 144:151–352.

Simberloff, D. 1976. Species turnover and equilibrium island biogeography. *Science* 194:572–578.

——— 1980. A succession of paradigms in ecology: essentialism to materialism and probabilism. *Synthese* 43:3–39.

Singh, R. S. 1979. Genetic heterogeneity within electrophoretic "alleles" and the pattern of variation among loci in *Drosophila pseudoobscura*. *Genetics* 93:997–1015.

Singh, R. S., R. C. Lewontin, and A. A. Felton. 1976. Genetic heterogeneity within electrophoretic "alleles" of xanthine dehydrogenase in *Drosophila pseudoobscura*. *Genetics* 84:609–629.

Sket, B. 1981. Fauna of anchialine (coastal) cave waters, its origin and importance. *Proceedings of the Eighth International Congress of Speleology, Bowling Green, Kentucky* 2:645–647.

Slatkin, M. 1974. Competition and regional coexistence. *Ecology* 55:128–134.

——— 1980. Ecological character displacement. *Ecology* 61:163–177.

Soulé, M. 1976. Allozyme variation: its determinants in space and time. In *Molecular evolution*, ed. F. J. Ayala, pp. 60–77. Sunderland, Mass.: Sinauer Associates.

Stearns, S. C. 1977. The evolution of life history traits: a critique of the theory and a review of the data. *Annual Review of Ecology and Systematics* 8:145–171.

Steeves, H. R. III. 1969. The origin and affinities of troglobitic asellids of the southern Appalachians. In *The distributional history of the biota of the southern Appalachians. Part I. Invertebrates*, ed. P. C. Holt, pp. 51–66. Blacksburg, Va.: Virginia Polytechnic Institute Press.

Strong, D. R. 1980. Null hypotheses in ecology. *Synthese* 43:271–285.

Strong, D. R., L. A. Szyska, and D. S. Simberloff. 1979. Tests of community-wide character displacement against null hypotheses. *Evolution* 33:897–913.

Sugihara, G. 1981. $S = CA^z$, $z \simeq 1/4$: a reply to Connor and McCoy. *American Naturalist* 117:790–793.

Sweeting, M.M. 1973. *Karst landforms*. New York: Columbia University Press.

Swofford, D. L., B. A. Branson, and G. A. Sievert. 1980. Genetic differentiation of cavefish populations. *Isozyme Bulletin* 13:109–110.

Tanner, J. T. 1966. Effects of population density on growth rates of animal populations. *Ecology* 47:733–745.

Thompson, P., H. P. Schwartz, and D. C. Ford. 1974. Continental Pleistocene climatic variations from speleothem age and isotopic data. *Science* 184:893–895.

Turanchik, E. J., and T. C. Kane. 1979. Ecological genetics of the cave beetle *Neaphaenops tellkampfi* (Coleoptera: Carabidae). *Oecologia* 44:63–67.

Turquin, M. J. 1981. The tactics of dispersal of two species of *Niphargus* (perennial, troglobitic Amphipoda). *Proceedings of the Eighth International Congress of Speleology, Bowling Green, Kentucky* 1:353–355.

Vandel, A. 1960. Les espèces d'*Androniscus* Verhoeff 1908 appartenant au sous-genre *Dentigeroniscus* Arcangeli 1940 (Crustacés; Isopodes terrestres). *Annales de Spéléologie* 15:553–584.

——— 1964. *Biospeologie: la biologie des animaux cavernicoles*. Paris: Gauthier-Villars.

Vandermeer, J. H. 1969. The competitive structure of communities: an experimental approach with Protozoa. *Ecology* 50:362–371.

Van Valen, L. 1975. Some aspects of mathematical ecology. *Evolutionary Theory* 1:91–96.

Van Zant, T., T. L. Poulson, and T. C. Kane. 1978. Body-size differences in carabid cave beetles. *American Naturalist* 112:229–234.

Vuilleumier, F. 1973. Insular biogeography in continental regions. II. Cave faunas from Tessin, southern Switzerland. *Systematic Zoology* 22:64–76.

Wake, D. B. 1966. Comparative osteology and evolution of the lungless salamanders, family Plethodontidae. *Memoirs of the Southern California Academy of Sciences* no. 4.

White, E. L., and B. M. Reich. 1970. Behavior of annual floods in limestone basins in Pennsylvania. *Journal of Hydrology* 10:193–198.

Wilbur, H. M., and J. P. Collins. 1973. Ecological aspects of amphibian metamorphosis. *Science* 182:1305–1314.

Wilkens, H. 1970. Beiträge zur Degeneration des Melaninpigments bei cavernicolen Sippen des *Astyanax mexicanus* (Filippi) (Characidae, Pisces). *Zeitschrift für zoologische Systematik und Evolutionforschung* 8:1–47.

——— 1971. Genetic interpretation of regressive evolutionary processes: studies on hybrid eyes of two *Astyanax* cave populations (Characidae, Pisces). *Evolution* 25:530–544.

——— 1973. Über das phylogenetische Alter von Hohlentieren. Untersuchungen über die cavernicole Süsswasserfauna Yucatans. *Zeitschrift für zoologische Systematik und Evolutionforschung* 11:49–60.

——— 1976. Genotypic and phenotypic variability in cave animals. Studies on a phylogenetically young population of *Astyanax mexicanus* (Filippi) (Characidae, Pisces). *Annales de Spéléologie* 31:137–148.

——— 1980. Prinzipien der Manifestation polygener Systeme. *Zeitschrift für zoologische Systematik und Evolutionforschung* 12:103–111.

Wolvekamp, H. P., and T. H. Waterman. 1960. Respiration. In *The Physiology of Crustacea,* vol. I, ed. T. H. Waterman, pp. 35–100. New York: Academic Press.

Woods, L. P., and R. F. Inger. 1957. The cave, spring and swamp fishes of the family Amblyopsidae of central and eastern United States. *American Midland Naturalist* 58:232–256.

Wright, S. 1964. Pleiotropy in the evolution of structural reduction and of dominance. *American Naturalist* 98:65–70.

Wu, T. T., E. C. C. Lin, and S. Tanaka. 1968. Mutants of *Aerobacter aerogenes* capable of utilizing xylitol as a novel carbon source. *Journal of Bacteriology* 96:447–456.

Index